To Grandaughter
Kirsten A. Foutt
Happy Birthday
Ap 3, 2015

Foutt

PLOWSHARES AND SWORDS

Tennessee Farm Families Tell Civil War Stories

by

Caneta Skelley Hankins & Michael Thomas Gavin

A Publication of the Center for Historic Preservation at Middle Tennessee State University

This publication may not be reproduced in any form without the permission of the Center for Historic Preservation

ISBN 978-0-615-88255-0   Price $50.00

❧ CONTENTS ☙

Foreword .... *i*

Preface .... *ii*

County Map of Tennessee .... *iii*

The Tennessee Century Farms Program .... *iv*

Acknowledgements .... *v*

As the Armies Advanced 1

The Farm as Battleground 25

    Sacred Ground 34

The Hard Hand of War 43

    Loyalty Oaths 53

    Photography in the Civil War 59

War and Work Animals 65

The Struggle to Farm 85

    Saltpeter and Gunpowder Manufacturing 101

Freedom to Farm 113

Rebuilding the Farm 129

    Remembering Veterans 141

    Post-war County Formation 146

    The Tennessee Department of Agriculture 154

Bibliographical Essay 162

About the Authors 164

Index 166

PLOWSHARES AND SWORDS

Tennessee Farm Families Tell Civil War Stories

The news travelled fast. Tennessee had seceded from the United States and joined the Confederacy. War was imminent! By a vote of 105,000 to 47,000, after months of prolonged and passionate debate, Tennesseans ratified a resolution of independence on June 8, 1861. The thought and the reality excited some and terrified many. What would war mean for their families and their farms? With mountains to the east, the Mississippi River to the west, and railroads, waterways, and rich farmlands in between, Tennessee was a vulnerable place and a strategic prize for whichever army could gain the advantage. The armies would come, but when or where no one knew.

With spring planting done, men soon began to enlist and drill with their chosen side, hoping to be back to harvest their crops and tend their livestock by fall. What happened to farm families and their property during the long years of war and afterwards, as told by the people themselves, by their descendants, and other sources, brings an acutely personal perspective to the Civil War and Reconstruction in Tennessee.

From the time the armies began spreading across the state in 1861, farm families were in the midst of fighting, foraging, and all the horrors of war. With a will to persevere and return to farming once the fighting ceased, veterans, widows, and their children, along with the newly freed men and women, began to build new lives, reclaim and revitalize the land, and apply progressive methods to improve livestock and crops. The legacy these families left, through their own stories, is vital to the interpretation of a time that remains the watershed of Tennessee history.

The Tennessee State Capitol opened in 1859. This view, with covered cannon on the steps, was photographed in 1864 by George N. Barnard.
Courtesy *Library of Congress*

⊰ PREFACE ⊱

"If history were taught in the form of stories, it would never be forgotten."

Rudyard Kipling, *The Collected Works*

Tennesseans are storytellers, and they enjoy hearing and reading stories about the rich and colorful history of the state and its people. Farm families, particularly those who have tended the same land for generations, are among the primary keepers of stories, photographs, and memorabilia. Their memories, documents, and records add perspective and understanding to each period of the state's history, but especially to the years of the Civil War and Reconstruction.

As the 150th anniversary of the Civil War approached in 2011, the staff of the Center for Historic Preservation considered ways to interpret the reality and the legacy of that time of trouble and transition. Since the late 1990s, the Center has administered the Tennessee Civil War National Heritage Area, a partnership unit of the National Park Service, and has been closely involved in planning and implementing programs and activities leading up to and in association with the sesquicentennial. Given the array of projects already underway with many excellent partners across the state, what could the Center provide to further illustrate this interval of challenge and change?

A publication based primarily on the stories of farm families was the obvious answer. Since 1984, the Center has coordinated the Tennessee Century Farms statewide program which recognizes and documents working farms that have been in the same family for at least 100 years. Files of farms in operation during or founded just after the Civil War were reviewed by Center staff who selected some of the most compelling stories. In addition to stories from certified Century Farms, accounts were taken from farms listed in the National Register of Historic Places; most of the nominations were prepared by Center staff and students. The farms finally chosen represent large and small acreages across the state, and both Union and Confederate allegiances. To illustrate the impact of the conflict geographically, counties in all regions of the state are included. For example, the first chapter, "As the Armies Advanced," begins in upper East Tennessee and concludes in Shelby County in West Tennessee. Other chapters start with farm accounts in Middle or West Tennessee, then describe the effects of the war on families in other parts of the state.

To support and expand the stories, additional research yielded facts, details, and context for the events and circumstances that occurred on the farms. A "Bibliographical Essay" describes some of the primary and secondary materials consulted in the preparation of this book, while information specific to individual farms is noted at the end of each chapter. Informational insets are included in several of the chapters to clarify topics that are mentioned in one or more stories.

The majority of photographs, historic and contemporary, were supplied by the farm families. Others are from the collection of the Center for Historic Preservation or were made for this publication by the authors. Contemporary sketches are reproduced from the following: Alfred H. Guernsey and Henry M. Alden, *Harper's Pictorial History of the Civil War* (Chicago: Star Publishing Company, 1896), which is abbreviated to *Harper's*; Louis Shepheard Moat, ed. *Frank Leslie's Illustrated Famous Leaders and Battle Scenes of the Civil War* (NY: Mrs. Frank Leslie, 1896), which is abbreviated to *Leslie's*; Paul F. Mottelay and T. Campbell Copeland, eds. *The Soldier in Our Civil War: A Pictorial History of the Conflict*, 1861-1865, 2 vols. (NY: Stanley Bradley Publishing Company, 1890), which is abbreviated to *The Soldier in Our Civil War*; and J. T. Trowbridge, *A Picture of the Desolated States; and the Work of Restoration* (Hartford, CT: L. Stebbins, 1868). Unless otherwise noted, maps are from *The Official Military Atlas of the Civil War*, Major George B. Davis, U. S. Army, Leslie J. Perry, Civilian Expert, Joseph W. Kirkley, Civilian Expert, compiled by Capt. Calvin D. Cowles, 23rd U. S. Infantry (Washington: Government Printing Office, 1890-1895). Otherwise, the source of each image is credited as appropriate.

The military conflict is designated throughout the book as the "Civil War" and the opposing armies and governments as "Federal" or "Union," and "Confederate" or "Secessionist." The exception is when other terms are used in direct quotations. Unless otherwise noted or cited, quotations are from Century Farm applications or nomination forms to the National Register of Historic Places.

County Map of Tennessee

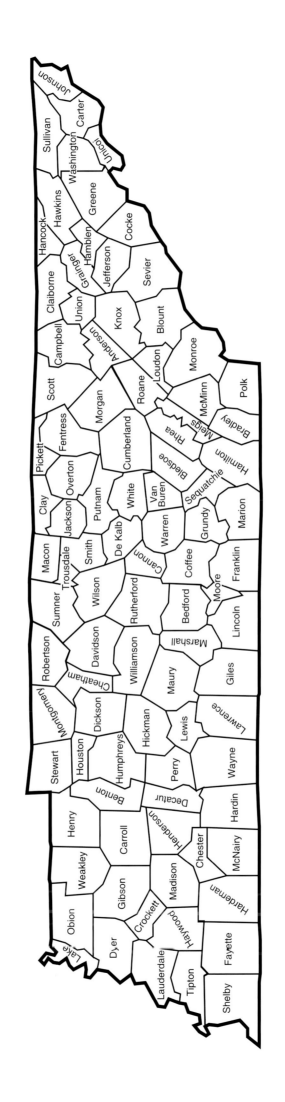

TENNESSEE CENTURY FARMS

The Tennessee Century Farms program recognizes and documents farms that have been owned and operated by the same family for at least 100 years. Initiated by the Tennessee Department of Agriculture as a national Bicentennial project in 1975, the program was transferred to the Center for Historic Preservation at Middle Tennessee State University in 1984. Since that time, the program has supported many projects. *Tennessee Agriculture: A Century Farms Perspective,* by Carroll Van West, was published by the Center and the Tennessee Department of Agriculture in 1986. A Century Farms exhibit, curated by Caneta Skelley Hankins, travelled the state from 1980-1981. Several counties have used the files to publish their own guides to local historic farms and agriculture, and many county fairs regularly recognize Century Farm families. An extensive Web site was created in 2008 which offers updated information on certified farms and the program. For nearly three decades, Center staff and students have recognized the collection's rich contents and produced an array of publications and studies, including nominations to the National Register of Historic Places, thesis topics, and scores of historic assessments.

Applicants to the program must prove that the farm has been in their family and in continuous agricultural production for at least 100 years. One owner must be a resident of Tennessee, and at least $1000 of annual farm income is required. As part of the application process, families are asked to include information about each generation of owners, as well as crops and livestock grown over the years. They are encouraged to submit copies of photographs of people, buildings, the landscape, and animals. These histories showcase activities and contributions, sometimes extending far beyond the community or county. As a whole, the Tennessee Century Farms collection is a significant repository of information covering more than 200 years of agriculture. Some properties, designated as Pioneer Century Farms, were founded before or in 1796 when Tennessee became a state. Even as this book was in production, however, it was learned that a few of the selected farms were recently sold out of the family.

To learn more about the Tennessee Century Farms program or to read the histories of certified farms, go to www.tncenturyfarms.org, or contact the Center for Historic Preservation, Box 80, Middle Tennessee State University, Murfreesboro, TN 37132, (615) 898-2947.

ACKNOWLEDGEMENTS

A project of this scope necessarily requires the willing assistance of many people, and to each person who contributed to this book, I am most grateful. Farm families generously shared additional details and images of their Civil War ancestors, as well as photographs of cemeteries and period farm buildings. Their cooperation and willingness to share their family stories brings their ancestors and their time into acute focus.

In several counties, local historians, archivists, librarians, educators, and University of Tennessee Extension Service personnel dealt efficiently with requests. Their interest in local farm stories and in the project as a whole is appreciated.

At the MTSU Center for Historic Preservation, Dr. Carroll Van West, director, and Dr. Antoinette van Zelm, historian for the Tennessee Civil War National Heritage Area, graciously carved time from their very full schedules to read this manuscript, and to apply their knowledge and editorial expertise to significantly improve this publication. Dr. West, director of the Century Farms program from 1985- 2001, is also the primary author of many of the National Register nominations cited. Center graduate assistants were enthusiastic participants and contributors to this project.

Leigh Ann Gardner, Center fellow and former Century Farms intern, helped in the initial review of farm files, and also compiled the inserts on "Loyalty Oaths," "Postwar County Formation," "Saltpeter and Gunpowder," "Photography," and "Remembering Veterans." Cassandra Bennett, Century Farms intern, also assisted in selecting some of the best stories, scanned photographs, and helped with specific research questions. Jaime Woodcock reviewed files and suggested initial farm selections. Julie Warwick contributed research from the agricultural census records, and diligently worked with historic images. Cynthia Duke, executive aide, and Debbie Sager, secretary, provided assistance in many ways.

Michael Thomas Gavin, co-author, passed away in January 2013, after fighting his personal war with courage and dignity. Mike was the preservation specialist for the Tennessee Civil War National Heritage Area at the Center for more than a decade. Along with a wealth of knowledge of the period, places, literature, and resource materials, Mike's scholarship, research and writing abilities, as well as his unwavering enthusiasm for this project were invaluable. His absence was felt keenly each day after his leave-taking. I shall always be grateful to Mike for his friendship, his dedication to history, and his essential contributions to this book.

Caneta Skelley Hankins

As the Armies Advanced

When the war became a reality, Tennessee farms lay unprotected and exposed in the path of both marching armies and the countless groups of bushwhackers who roamed the state. For some families, especially those who lived and worked along main roads and rivers or whose property was near strategic points on the landscape, soldiers within and around their houses, outbuildings, and fields became commonplace as they endured the war years.

Across the Volunteer State, partisan allegiances were complex, and both the Union and the Confederacy retained support in each of the Grand Divisions. For perspective, in February 1861, the voting men of Tennessee had soundly rejected a referendum calling for a convention that would determine whether or not Tennessee should secede from the Union. By June of the same year, a resolution of independence was ratified by a vote of 105,000 to 47,000. This dramatic change has been attributed to both President Abraham Lincoln's call for troops following the firing on Fort Sumter in April and to the intimidation of Unionist voters. Though the final vote swung to secession and joining the Confederacy, pro-Unionists could be found in virtually every county. Likewise, in those areas that remained primarily Union, Confederate support was not completely unknown.

In East Tennessee, the landscape of challenging mountain slopes drained by rivers flowing through fertile valleys was one of the state's richest agricultural regions in the mid-nineteenth century. After the results of the referendum of February 1861, the majority of voters there had assumed that the state would remain in the Union, but by late fall they found themselves included in the new Confederacy. Most residents, however, remained overwhelmingly loyal to the government of the United States, though some families, or members within a family, were sympathetic to the Confederate cause. These individual and collective decisions and their consequences affected the region for generations.

Middle Tennesseans were also divided over secession. Two-thirds of all free families in these counties derived a part of their income from farming or agricultural-related business in 1860. Farms tended to have small acreages of which more than half were deemed less than $2,500 in value. Farms with more than twenty slaves accounted for just seven percent of the total, but these planters had considerable influence in the state and the southern region. Although connected to the commercial centers of the Midwest, the more successful farmers in the fertile valleys were likely to support the Confederacy, while the small farmers and stock raisers in the hills and hollows surrounding the Central Basin tended to remain loyal

Farms were regularly visited by soldiers who were constantly moving across Tennessee.
From *The Soldier in Our Civil War*

Farm families often found soldiers in their barns and fields, and rail fences were prized because they were easy to disassemble to use for firewood. From Leslie's

to the Union. The military activity that occurred in Middle Tennessee ranged from nearly continual troop movement to major battles, beginning with Ft. Donelson in early 1862 and concluding with the battles of Franklin and Nashville at the end of 1864. Differing allegiances in the heartland led to disagreements, grievances, and violence that rarely ended with the Confederate surrender at Appomattox.

The hot, humid climate and large flat tracts of land of the Mississippi Delta in West Tennessee made a perfect environment for cotton, wheat, and corn. Control of the Mississippi River and Memphis, the state's largest city and the cultural and economic center of the regional cotton empire, were prizes the Union claimed early. Although the slave-holding agriculturists in this part of the state adamantly supported secession and voted consistently for that option, pockets of support for the former government existed, particularly along the Tennessee River. The control of the vital rails and rivers in West Tennessee was continually contested as troops traversed the region.

What might happen when soldiers came to their farms was an ever-present fear of men, women, and children in the countryside. The accounts of what did occur demonstrate how those at home, no matter their allegiances, endured the intimidating presence of the soldiers who advanced across every section of the state during the war years.

Still Hollow Farm

Greene County

Still Hollow's rich bottom lands have always been prime farming acreage, and Allen's bridge spanning the Nolichucky River held strategic interest for troops. The stones from the Civil War bridge were used to construct this wall when the bridge was replaced years later.

The farmhouse at Still Hollow Farm was built ca. 1861. Other buildings that remain from the Civil War period include a granary, log portions of the dairy and tobacco barns, and a log smokehouse.

Greene County, home of Tennessee's Civil War military governor, Andrew Johnson, is typical of East Tennessee farming communities that dealt with Union and Confederate troops. In 1857 James Allen, Sr., and his wife, Laura M. Brown Allen, acquired property now known as Still Hollow Farm. Here they raised corn, wheat, horses, cattle, and hogs. The Allens were a respected and influential family who, according to the 1860 federal census, owned land and slaves worth more than $21,000. When the conflict began, the Allens supported the Confederacy, although they lived in a predominately Unionist county.

When a new bridge was completed across the Nolichucky River at the Allen farm about 1862, the span drew the attention of military authorities. Allen had a reputation as a "rich rebel," who supported the Confederacy and probably aided the partisan bands still active along the Nolichucky late in the war. In April 1865, a Federal force went to Allen's bridge searching for a notorious gang of guerillas.

Gen. Washington L. Elliott reported, "I sent the Twenty-fourth Wisconsin Volunteer Infantry, Maj. MacArthur [Arthur MacArthur, the father of Gen. Douglas MacArthur of World War II] commanding, accompanied by a sergeant and twelve men of the Eighth Tennessee Cavalry, familiar with the country and people, to Johnston's and Allen's Bridge, over the Chucky. The major has reported that five guerrillas of Tully's band, from Hamilton, Cocke County, Tenn., were at the bridge on Friday last." On this scouting mission, however, the Federals found nothing.

Another Civil War-era story regarding the Allen family, which can only be verified in part, is that in late 1864, James Allen discovered that two furloughed Federal soldiers, William P. Seaton and John Davis, had slipped back home to visit their pregnant wives. Both men rode with the 8th Tennessee Cavalry (USA). Supposedly, Allen informed Confederate authorities of their whereabouts, and soon afterwards the two Union cavalrymen were murdered in nearby Parrottsville.

After the war, James Allen, Sr., remained in Greene County until his death in 1885. He continued to farm his land successfully, raising corn, wheat, and livestock. Although the loss of his slaves affected his finances considerably, the value of his real estate had doubled by 1870. Still Hollow, also known as the Allen-Birdwell Farm, is listed in the National Register of Historic Places.

Fermanagh-Ross Farm
Greene County

Families from Ireland were living and farming land near the Old Ridge Road, also known as Ross Road (now Tennessee Highway 93), by the late 1700s. William Ross, of County Fermanagh, was a landholder in Greene County as early as the 1780s. In 1813, his son, William Ross ll, married Margaret "Peggy" Gass, whose parents had also emigrated from Ireland. This first-generation Irish-American couple built a substantial brick house, Maden Hall, in the mid-1820s.

As Ross improved the farm on which he grew a variety of row crops and livestock, he constructed several outbuildings, including a log smokehouse, log corn crib, a mill, a cantilever barn, and a detached kitchen/slave cabin. Ross farmed with the help of several slaves. He inherited at least four slaves from his father in 1834 -- a boy Jess, a woman and her child Mary, and Jack, an adult male. In 1860, twelve slaves were listed on the farm along with one identified slave cabin.

John Gass Ross, grandson of William and Peggy, wrote a memoir based on his days on the farm during the Civil War. A young boy at the time, he recalled the names of slaves, including "Aunt Sarah" and her son, Louis, as well as "Old Barney," and also "Little Jim," with whom he played in the cantilever barn. He reported that the slaves played music and were fond of dancing, and that they attended the local Presbyterian Church. His schoolteacher, Bell Wilson, "taught in one of the negro houses on Grandfather's place." She apparently taught either the enslaved people or post-war emancipated blacks to read and write.

Ross, who lived in Rheatown, also remembered that his grandfather invited all of his family members to come live with him at Maden Hall during the war because he felt they would be safer there than at their individual homes. The farm itself was self-sustaining and the elder Ross was sure his family would always have plenty to eat. In time, however, the farm was raided not only by uniformed soldiers (who, Ross noted, took only necessities, usually food), but also by bushwhackers who stole everything they could carry.

Ross recalled that an old mare and a young mule colt were hidden and locked in the log smokehouse, and chickens were concealed beneath the floor of the kitchen. Furniture was stored in two locked upstairs bedrooms, but even that did not deter bushwhackers from stealing clothing and furnishings, as well as tools and plows and every animal they could find.

Toward the end of the war, William and Peggy Ross moved to southwest Virginia where they both died in 1865. Their son, William Ross III, a Unionist, filed extensive and substantial claims with the Southern Claims Commission trying to recoup some the family's losses from the U. S. military. He gave evidence of unwavering support of the Union by explaining that he has been captured in late 1861 or early 1862 and taken to Greeneville where he was impressed to take the Confederate Oath at the point of a bayonet. A neighbor, Robert Carter, verified that Ross, too old to serve at 48, had volunteered to join Company A of the East Tennessee Mounted Infantry (USA) in 1864. He was prevented from doing so when captured again by Confederates as he travelled to Strawberry Plains where the group was to muster.

Ross's claims were unsuccessful, and the family struggled with debts and restoring the farm's fields and stock during the years following the war. William and his wife and four grown children, however, made some progress. By 1870, the farm was valued at $4,500 and personal property totaled $2,500. William Ross III died in 1893, leaving the farm and Maden Hall to his son, Vincent, who survived his father by only four years. His widow, Mary Elizabeth "Mollie," who had lived on the farm since 1879, took over its operation and continued to manage it well into the twentieth century.

The slave house/kitchen (in foreground) was located to more easily serve the occupants of Maden Hall.

Wagner-Worley Farm
Johnson County

The Wagner-Worley Farm has several buildings and sites that remain from the period before the Civil War came to upper East Tennessee. The founder, David Wagner, settled and established the farm in 1790. He fathered seven children with his first wife, Mary Catherine Hagey, and then ten more with Margaret Peggy Weitzel. Nathaniel T. Wagner, the fifth child of David and Margaret, built his home ca.1850 when he inherited the farm of about 200 acres on Roan Creek. Nathaniel was married three times and fathered sixteen children. Johnson County residents were generally sympathetic to the Union, but the Wagners, of which there were many because of the large families of the first and second generations, were mainly pro-Confederate. In 1860, Margaret Wagner owned two slaves, a female aged 90 and a male who was 70, but Nathaniel's name does not appear as a slaveholder in that census. The farm was a diverse operation producing corn, barley, flax, sugar cane, oats, wheat, hay, and tobacco along with cattle, swine, and sheep.

The Wagners were raided on several occasions, and the family learned to hide corn in stacks of firewood. One of the chores given to the children was to go out each morning and pick up any kernels of corn that had been pulled out during the night by small animals. This evidence would have revealed the staple grain's hiding place to foragers.

What remains from the farm's early settlement and antebellum periods is a log barn built ca. 1790-1810, portions of the 1850 house, a well and well-house which have changed little since 1850, and an apple house where apples and root vegetables were stored in an area that was "dug into the ground and insulated well with sawdust." Built above the apple house is a two-story smokehouse; the second level was used for a weaving house. The loom there was apparently assembled sometime in the period before the Civil War. Tom and Mary Ann Worley, the current owners, manage and operate this family farm that pre-dates Tennessee statehood.

The current farmhouse retains the core of this house, built in 1850 by Nathaniel Wagner, within the remodeled residence.

The Wagners sold corn and hay to Confederate troops on December 29, 1864.

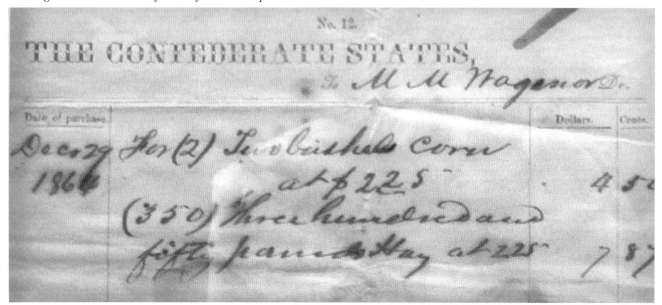

Lamar Farm
Anderson County

The story of the Lamar Farm, founded in 1851, illustrates the frustration and alienation that occurred within immediate and extended families as they coped with the realities of war. Rose Lamar Longmire, who applied for Century Farm designation in 1975, was most knowledgeable about her family's heritage and was a keeper of oral history from the period. She explained that she "vividly remembered" her grandmother, Nancy Wallace Lamar, who lived to be ninety-three years old, telling stories about the Civil War. Her grandfather, Joseph B. Lamar was too old to serve, but his eldest son, William, was mustered into the 2nd Tennessee Infantry (USA) at Clinton when he was seventeen. Outraged by this action, another son left home and joined the Confederate army.

One night when the family was asleep, a knock was heard on their door and a Federal officer requested a "bed and a good breakfast." The family was surprised to see their long front porch covered with sleeping soldiers. Before the troops moved on, they burned the split rail fences for their campfires, collected wagonloads of corn for fodder, and seized all the good horses, leaving their worn-out and injured mounts behind. The soldiers also gathered up all the meat, chickens, flour, meal, and livestock that they could find. One daughter managed to save a sack of cornmeal by sitting on it and hiding it with her long, full skirt. As their situation became even more difficult, Nancy Lamar rode into Kentucky to a trading post to purchase supplies. She was detained, questioned, and searched by Federal soldiers who even took "off her shoes." When she explained to them that she had a "boy in that army," they allowed her to continue her journey.

As further illustration of the complexity of family relationships and the sometimes incongruous perspectives of individuals during this period, Rose Lamar Longmire's maternal great-grandfather was U.S. Gen. James G. Spears of Bledsoe County. Although a slaveholder, he strongly supported the Union. When he learned that he was to be arrested for his loyalty, he went to Kentucky, where he assisted in organizing the 1st Tennessee Infantry Regiment (USA). By the following March he was promoted to brigadier general. Spears believed

Like other women, Nancy Lamar, travelled behind Union lines to buy or beg for food for her family.
From *The Soldier in Our Civil War*

that the Emancipation Proclamation was neither legal nor constitutional because it illegally deprived him of his slaves, and, despite his high rank, he persistently expressed his opinion (Spears's active opposition is somewhat ironic given that Tennessee was excluded from the Emancipation Proclamation specifically to avoid alienating Unionist slaveholders like him). In February 1864, President Abraham Lincoln authorized a full investigation which resulted in Spears's court-martial and dismissal from the army the following August. Though this matter did not directly affect the farm life of the Lamars in Anderson County, it was a topic of family discussion for generations and serves as a reminder of the many paradoxes of war.

Through stories told by Nancy Wallace Lamar, her descendants understand how the family coped during troubling and difficult years

Tarwater Farm
Sevier County

Matthew Tarwater was photographed not long after the Civil War.

In the mountainous terrain of Sevier County, Matthew Tarwater, of Dutch and German ancestry, and his wife, Sarah Rule, began to cultivate their land in the late 1850s. They had a diverse farm of row crops, livestock, and poultry. The industrious family lived in a sturdy house, and the various outbuildings included a smokehouse and large two-story granary that was well stocked with grains as well as nuts and fruit. The Tarwaters were also known for the quality of their Irish potatoes. Matthew Tarwater was a stockholder in the East Tennessee & Georgia Railroad.

Sevier County was strongly pro-Union during the Civil War, and most local men eventually joined a Federal army regiment. Because of the sentiments of the county's residents, Confederate raids were frequent and destructive. Time and again, foraging troops "robbed the family of food, provisions, horses, and cattle," and on one occasion also "a considerable sum of money," as recorded in *A Brief History of the Tarwater Family*, by Charles B. Tarwater. He wrote that the family, which in the 1860 census included seven children under twelve years of age, also "suffered much personal abuse." To cope with the raids, food was stored in caves and under piles of brush, but often the "intruders would watch until they found out where these provisions were hidden, and then take them." The farm's current owner, Doris Tarwater Phelps, can pinpoint a "narrow hollow on the back of the farm" where her father recounted that Matthew "hid his hogs from Confederate soldiers."

After the war concluded, family members, now seven sons and two daughters, were active in rebuilding the county. Matthew served on the Sevier County Court, and he and his sons helped to construct the Pleasant Hill Methodist Church, where they were charter members. Both Sarah and Matthew survived the war by several years and, in 1898, Matthew started an annual family Thanksgiving reunion that is held each year. The founders' graves are marked in the Tarwater Family Cemetery.

The family's house and detached kitchen are the backdrop, in 1897, for three generations of Tarwaters. Matthew is seated between his grandsons, Freeman and Millard. Standing behind him, from left, are his daughter, Sarah Elizabeth, and his son, Adam, whose wife is holding their daughter, Gertrude.

Fowler-Lenoir Farm and J & J Farm
Monroe County

From Monroe County come stories from two families who coped with advancing troops in different ways. Charles and Elizabeth Wyley Kelso, established their homestead on Fork Creek in 1824, and he was a commissioner of Madisonville. Their daughter, Mary Josephine Kelso Fowler, received her portion of the farm in 1853. Her husband, William J. Fowler, was a progressive farmer and stockman who also operated a grist mill and a saw mill in the Eve Mills community. The farm is situated along the Sweetwater Road and Fork Creek which empties into the Little Tennessee River about four miles away. This location put the family and the farm in jeopardy.

During the Civil War, Federal soldiers took food and supplies from the farm on several occasions. One of these visits resulted in William Fowler being threatened with hanging because of his Confederate sympathies. James Wyley Kelso, his wife's older brother and a Unionist, intervened and persuaded the soldiers to leave him alone.

In front of the house (no longer extant) that sheltered the family during the Civil War is Elizabeth Wyley Kelso, her daughter, Mary, her son-in-law, William Fowler, and her granddaughter, Cora Fowler, along with the family dog.

The post-and-beam barn on the Fowler-Lenoir Farm dates from the late 1850s. The "Yankee bench" survives as part of the Civil War lore of the family.

The family remembers that Gen. William T. Sherman's troops, as they prepared to cross the Little Tennessee River at Morganton in 1863, camped at the farm. At times, while marching past the house, footsore soldiers would stop and rest on a wooden bench in the yard; it has been called the "Yankee bench" by each generation of the family since that time.

In the 1880s, Fowler was involved in the formative years of the Farmers' Alliance, an organization which urged farmers to be more aware of local, state, and national issues and to adopt scientific methods, diversify their crops and livestock, and become as self-sufficient as possible to reduce their debt. Fowler was a staunch Democrat who served two terms (separated by a decade) in the post-bellum Tennessee legislature.

The experience of Joseph Ragon, also of Monroe County, adds another dimension to the way families coped with the war that raged around them. Ragon, who owned 360 acres, now the J & J Farm, refused to fight for either side. He built a lean-to in the dense woods and hid out for the duration of the war. On several occasions he could hear the sounds of battle not far from his shelter. His wife, Jane Lilliard Ragon, and their children stayed on the farmstead during this period and brought him food. Ragon did not want to shoot any game for fear of disclosing his hiding place. The Ragons' son, Joseph Charles, who owned the land until his death in 1954, passed along the story to his descendants who still live and farm the same land.

Joseph Charles Ragon, left, along with his son, Horace, center, and an unidentified man are at the house where the family lived during the war and for some years afterward.

Teague Farm
Marion County

When the Civil War began, Elijah Teague owned and operated a farm near Walden's Ridge, which was bounded by the Sequatchie River for over a mile. Teague was listed as a "distiller and rectifier," which meant he not only made whiskey but also blended it. Living along the Signal Mountain Road and only 20 miles from Chattanooga was good for business, but Elijah and his wife, Rebecca Looney, also saw a lot of military activity during the war. Soldiers on the move would camp near the river, creeks, or the household spring at any time of day or night. The Teagues have passed down several family stories from the war years.

One tale involves a young Joel E. Teague who would visit the soldiers, listen to their talk, and beg for sugar. Sugar, among other basics like salt and flour, was in short supply or non-existent for southerners as the war years extended. Another memory features the recovery of a discarded checkerboard that was used for many years "as a dough board" by

A raid on a Federal supply train by Confederate Gen. Joseph Wheeler took place near the Teague Farm in 1863.
From *The Soldier in Our Civil War*

the family after the soldiers left the farm. A third recalls a raid in early October 1863, when Gen. Joseph Wheeler's cavalrymen attacked and burned a large Chattanooga-bound Federal supply train a few miles from the farm. Soldiers killed hundreds of mules and destroyed an enormous amount of military equipment and supplies. Robert and Joel Looney, Rebecca Teague's brothers who lived nearby, went to the scene of the raid and collected scraps of lead from the wreckage to mold into bullets. They later enlisted and fought in the Union army. Although the senior Teague was never a soldier, he was shot in the shoulder by a bushwhacker and was bedridden for weeks recovering from the wound.

Besides its proximity to the Sequatchie River, the Teague Farm was adjacent to the main road from Walden's Ridge to Chattanooga. For a time it was the only supply route for the Union army and cavalry skirmishes occurred frequently. Its turnpike location caused the Teague Farm to be raided for food again and again. The women became tired of the constant theft and took action to prevent more loss. Rebecca hid twelve hens and a rooster under the floor of the house. If the rooster would start to crow, Rebecca would "whack him with a broom" to quiet him. Elijah's sister, Margaret, once faced off soldiers who were about to take their last side of meat by brandishing an ax and threatening, "Leave that alone or I'll chop you in two." The soldiers left empty-handed.

After the war, Joel Teague (the same youngster who visited the soldiers) became the owner of the farm. He practiced progressive farming methods and was active in rebuilding and improving the community and county. He owned and operated a blacksmith shop and installed the only set of scales in the area for weighing wagons and livestock. Teague also advocated for the construction of the first modern paved road over Walden's Ridge to provide easier access between the Sequatchie Valley and Chattanooga.

McPeak Farm
White County

In 1863, a White County farmhouse was spared just as it was about to be burned by Union soldiers. The story involves the original homesteader, David Clark, who moved onto the land in the early 1840s and built a frame house in 1858 just off the Old Kentucky Road, a main route from Maysville, Kentucky to Alabama. With the outbreak of the war, the two older Clark sons, Phineas and Darius, who lived at home and helped their father farm, joined with other local men to form Co. G of the 16th Tennessee Infantry (CSA). Darius Clark kept a diary during his service. On September 9, 1862, Sgt. Clark wrote that his company, having travelled north from Mobile, Alabama, arrived in Sparta, the county seat, on September 3, and "our father and sister, and brother came to see us which gave us grate pleasure and a grate many of my kinsfolks came up to see us all." The final entry in his diary, dated October 7, 1862, reads, "We are lying in the sun waiting for orders to move on." Darius Clark died early the next morning during the Battle of Perryville, Kentucky.

In 1863, still grieving for Darius, David Clark, nevertheless, prevented Confederate sympathizers from attacking a lone Union soldier who was found in a nearby community. Sometime afterwards, Federal troops were reported to be marching through the area, burning dwellings and barns in their path. David Clark and his family took their belongings to a "swampy sage field" behind the house. The troops arrived with torches, but a lone soldier rode forward and requested that the group move on without setting fire to the farmhouse and outbuildings. Clark recognized the man as the one he had saved from a brutal beating. The column passed by the Clark homestead and left the property unharmed.

After the war, David Clark moved to northern Alabama, and the property was divided among his family. His daughter, Mahaley, who married John Lambert McPeak, received a portion and the house. The McPeak family continues to live in the original Clark house.

Mahaley Clark McPeak, center, is with her family and their horses, ca. 1880s, in front of the house that was spared because her father, David Clark, rescued a Union soldier from a beating.

Lairdland Farm
Giles County

The almost endless troop movements throughout the war years continuously exposed the soldiers to unfamiliar people, places, and things. Lairdland Farm, located in Giles County's Brick Church community about ten miles northeast of the county seat of Pulaski, was a large antebellum plantation of 1,063 acres near the turnpike that connected Pulaski and Lewisburg. Robert and Nancy Laird lived the lifestyle of the planter class, centered on their 1831 Federal-style frame residence. The Lairds had one daughter, Mary who was called "Mackie," and managed a very successful large-scale farming operation, raising cotton, corn, mules, and cattle by means of slave labor. In the 1850 slave census, forty-nine men, women, and children were owned by Robert Laird. According to the 1860 federal census, the Lairds had accumulated almost $140,000 in real and personal property which included their slaves.

In the autumn of 1863, the Confederate Army of Tennessee's cavalry commander, Gen. Joseph Wheeler, led a successful raid on Federal supply lines in Middle Tennessee. But when Wheeler attempted to return to his base in Alabama, Federal forces under Gen. George Crook caught up with Wheeler's men near Farmington on October 7. Riding with the Confederate column that day was Capt. James K. P. Blackburn, leading a company of the 8th Texas Cavalry Regiment (Terry's Texas Rangers).

During the battle, Blackburn and three of his men were wounded severely and left in the care of local citizens. While recovering, Blackburn met and fell in love with Mackie Laird. He returned after the war and married her in February 1867 on the front porch of the farmhouse. Blackburn played a leading role in rebuilding Giles County and served in both houses of the state legislature. In 1918, Capt. Blackburn published a lively account of his wartime experiences entitled *Reminiscences of the Terry Rangers.*

The house, built by the Lairds in 1831, was the scene of the 1867 wedding of Mary M. (Mackie) Laird and Capt. James K. P. Blackburn.

Meadow Dale Farm and Wooten-Kimbro Farm
Bedford County

Two farms, both of which originated with Mary L. Elkins Kimbro, are part of the history of Normandy, one of scores of settlements that were created during the 1850s along the tracks of the main line of the Nashville & Chattanooga Railroad. During the Civil War, Normandy's strategic location on the railroad and the Duck River first brought Union soldiers to the area to guard the railroad bridge during the Tullahoma campaign of 1863. That summer, Gen. William S. Rosecrans (USA) led the Army of the Cumberland against forces under Gen. Braxton Bragg (CSA), driving the Confederates south, out of Middle Tennessee. Four forts were built to guard the N & C Railroad tracks and bridge. Small square enclosures, two forts were located on each side of the bridge about 50 feet from the trestle.

Mary L. Elkins Kimbro, one of the few female founders of a Century Farm, acquired her acreage in 1852. On Meadow Dale Farm, family documents indicate that about fifteen enslaved persons lived and worked. Mary's husband, Allen, died before the decade ended, leaving her to raise six sons, all under eighteen years of age. Her son, Thomas, a Confederate cavalryman, died during the war. Another son, George, was fifteen years old when he saw one of the Army of the Cumberland's commissary wagons pass by his house heading toward Chattanooga. He planned a raid that night to capture "pies and ginger cakes." His brother, Benjamin, tried to persuade him otherwise but, when he realized his brother was serious, told him to bring him some of whatever he managed to take. Family tradition holds that young George was successful in his mission and returned with the rare treats.

On another day, George was taking corn to the mill when a bullet whizzed by, narrowly missing him. When the Union soldier who fired the shot emerged from his hiding place and saw how young George was, he swore that he wasn't really shooting at him, but at a crow instead. As a token of good will, he gave George a "canteen and a musket which had a broken spring." Both of these artifacts remain in the family.

In 1910, George Ernest Kimbro bought land near his grandmother. This acreage is known today as the Wooten-Kimbro Farm. Remnants of two Civil War forts remained on the property when George purchased it. One has since been plowed though the other is still visible. The family reports finding minie-balls over the years as evidence of the fighting that took place for control of the rails over the river.

Mary Elkins Kimbro was among very few women who established a Century Farm before the Civil War.

Benjamin S. Kimbro was not quite as adventurous as his brother, George, but both remembered troops camping nearby.

The current railroad bridge that spans the Duck River at Normandy is believed to be set on the stone abutments and pillars of the Civil War-era bridge.

Vaughn Farm
Stewart County

In northern Middle Tennessee, the homestead established in 1838 by Virginians Joseph W. Gray and his wife, Rachel Fulkerson, in Stewart County is known as the Vaughn Farm. During the mid-nineteenth century, it was one of the largest holdings in the county and yielded corn, wheat, cotton, tobacco, cattle, and swine from well over 2,200 acres. After their wedding in 1842, the Grays' daughter, Nancy Emily, and her husband, William B. Weaks, received a portion of the productive farm and eight slaves from her parents as a gift.

In 1861, two sons, Andrew and James Weaks, joined the 50th Tennessee Infantry (CSA), which later fought at Fort Donelson. Their grandmother, Rachel Gray, who lay in bed dying of pneumonia in February 1862, and the rest of the family who had gathered in the farmhouse, could hear the musket fire, the rumble of heavy artillery, and the continuous maneuvers of the opposing armies during the Battle of Fort Donelson. After the fort's surrender, the brothers managed to escape, along with about half their regiment. They joined the 2nd Kentucky Cavalry, rode with Gen. Nathan B. Forrest for a year, and were part of Jefferson Davis's escort when he was captured at Irwinsville, Georgia, in May 1865.

Another of the Grays' sons, Frederick, purchased the Dover Hotel in July 1860. This building served as Confederate Gen. Simon Buckner's headquarters during the fighting. It is now known as the "Surrender House" because it was the site of the "unconditional and immediate surrender" of Buckner to Gen. Ulysses S. Grant on February 16, 1862. The building, restored by the National Park Service in the 1970s, is part of Fort Donelson National Military Park.

The Dover Hotel, also known as the "Surrender House" at Ft. Donelson, was owned by Frederick Gray.

Edgman Farm
Henry County

Henry County's eastern border is the Tennessee River, the natural boundary between the middle and west portions of the state. Seven miles southeast of the county seat of Paris near the Big Sandy River, brick mason John Payton Kimbrough and his wife, Nancy Higginson, founded a 150-acre homestead in 1850. Their daughter, Sarah, born in 1820, was their only surviving child. She married Jeremiah Boden, and when the war commenced, their son, James W. Boden, enlisted in the Confederate army's 5th Tennessee Infantry along with many other Henry County men. Pvt. Boden was once captured at Murfreesboro, but managed to escape and soldiered throughout the war.

In her later years, Sarah Boden well recalled those dreadful times and passed along her first-hand experiences to her grandchildren and great-grandchildren, including Aline Boden Love, who recorded the stories in a family scrapbook. One account described how the farmhouse was used as a Confederate hospital in 1862, and "they dug bullets" from many of the soldiers and dressed their wounds, "using all the sheets, pillow cases, and even their petty coats for bandages -- even to the curtains from their windows." Sarah also explained how her father died. When Federal gunboats began shelling inland from the Tennessee River, the east chimney of the house was hit and John Kimbrough "fell dead with heart failure." Neighbors made the coffin in which he was buried near the house.

Nancy Kimbrough, a "small but spunky woman," had no one to help her run the farm after her husband's death. On one occasion, she was warned that Union soldiers were approaching her home. Hoping to keep from being robbed, she "gathered an apron full of fresh baked tea cakes and went out to meet them." The soldiers gratefully accepted her peace offering and marched on without harming her or her property.

In order to assure the future of the farm, Nancy made an agreement with her grandson, James, that if he would take care of her and the land until she died, the property would become his. After he was discharged, he came home, accepted her offer, kept his promise, and inherited the house and land in 1875.

The original log dog-trot dwelling on the farm was built by J. P Kimbrough about 1830, and was used as a field hospital in 1862. The house was enclosed with weatherboarding in the early 1930s. Because of termites and deterioration, the house was torn down in the 1990s and a new house built on the site. Today, Kimbrough descendants call the property the Edgman Farm.

Keeton Farm
Decatur County

West of the Tennessee River lies Decatur County, where Robert and Catherine Keeton settled on slightly more than 157 acres in 1825. Their land was seven miles south of Scotts Hill near the intersection of the Old Stage Road and Dunbar Road. The parents of ten children, the Keetons grew corn, cotton, and hay and raised cattle. In addition to his agricultural pursuits, Keeton practiced medicine. Dr. Keeton died in 1858, leaving a successful farm and business, and is buried in the family cemetery with other relatives. His son, John Lawson Keeton, also a physician, inherited the farm.

The Civil War had an immediate impact on the Keeton Farm. A Confederate mustering station was set up near the busy crossroads, and men and boys from the vicinity came to enlist in the army. These Decatur County recruits joined established units such as Cox's Tennessee Cavalry Battalion or the 27th Tennessee Infantry Regiment. In mid-December 1862, some of these soldiers returned with Gen. Nathan B. Forrest when he crossed the Tennessee River at Clifton at the start of his first West Tennessee raid. After spending the better part of three days crossing the river, the weary brigade encamped in a grove of tall trees near Hermitage (present day Dunbar). The women and girls of the community had stayed up the previous night to prepare enough food for more than 1,800 Confederate troopers. Two weeks later, hotly pursued by a large Federal cavalry force, Forrest's command raced right by the Keeton Farm, heading for safety across the Tennessee River.

The Keeton Store at Dunbar dates from the 1840s when it was a tavern and stagecoach stop. It was a Confederate mustering station during the Civil War.

Country Wood Farm and Holt Farm
Gibson County

Hester Farm
Fayette County

The presence of both Union and Confederate forces was a regular occurrence on any farm that was adjacent to or near a railroad. For Jane Moore Dickson of Gibson County, who founded Country Wood Farm with her husband, Thomas, in 1848, the location south of Trenton near the tracks of the Mobile & Ohio Railroad was important for market and transportation. In 1860, the Dicksons owned seven slaves, possibly one family as the oldest female was twenty-eight, the oldest male thirty, and five children ranged in age from two to fourteen years. On nearly 300 acres, food was grown for the residents, and cattle and cotton were primary cash products. When the railroad was being built, Jane Dickson, by now a widow for several years, hired out slaves to help with the construction. By 1861, her daughter, Isabella Dickson Cooper, the widow of James Irvin Cooper, owned a portion of the land. The two oldest sons of Isabella and James served in the Confederate army. Isabella, Jane, and the younger children carried on with the farm work and became accustomed to military activities and their consequences during the war.

Likewise, movements of both armies affected the Holt Farm which was also located along the Mobile & Ohio Railroad. In December 1862, Gen. Nathan B. Forrest's cavalrymen passed through the farm and camped at the nearby trestle while they burned bridges and tore up the tracks.

To the south near the Mississippi border, the Memphis & Charleston Railroad ran through the Hester Farm, which was founded in Fayette County in 1854 by Charles Brame. Elizabeth Brame Hester and her husband, John, owned the property during the Civil War. After Union forces occupied Memphis in the summer of 1862, the Hesters became accustomed to soldiers moving supplies back and forth and, according to the family, "Federal troops made a practice of firing at the family dwelling" as the train passed.

Farms near railroads saw frequent military activity as troops either fought for control of the rail line or destroyed the tracks so the other army could not use them.
From *Leslie's*

Barret Farm
Shelby County

In 1850, Anthony Robert Barret purchased nearly 76 acres of Shelby County land from his brother-in-law, William A. Polk. Barret married Rebecca Hill, from Norfolk, Virginia, later that year, and the couple eventually raised four children on their farm. Barret also founded the village of Barretville. In 1856, he opened the first general store in the community which offered food, hardware, clothing, fertilizer, and various sundries; extended credit; provided mortuary services; and served as a gathering place where the local farmers and their families could socialize and share the news of the day.

Anthony Barret, 1827-1910, founded Barretville.

Barretville prospered until the disruption of the Civil War. After the surrender of Memphis in June 1862, most of West Tennessee fell under Federal control. That summer, local Confederate supporters Robert V. Richardson, Aaron A. Burrow, and John U. Green organized the 1st Tennessee Partisan Ranger Regiment behind the Union lines in Tipton and Fayette counties to harass the occupiers. Following the winter of 1862-1863, during which the Rangers fought a series of skirmishes with U.S. Col. Benjamin Grierson's cavalry, the Federal high command in Memphis ordered mounted columns sent from different points to converge on the lower Hatchie River area and search out and destroy these partisans.

Unable to withstand such formidable forces in a pitched battle, the regiment temporarily disbanded, and the soldiers spread out over the countryside. In April 1863, after eluding Union patrols in Big Creek Bottom, a small detachment including Col. Green, James M. Barret (Anthony's brother), and three other men stopped at the Barret Farm for supper and for fodder for their horses. Suddenly, Col. Fielding Hurst and his 6th Tennessee Cavalry (USA) appeared and captured all but one of the Confederates. Hurst's regiment of West Tennessee Unionists continuously scouted and patrolled the no-man's land beyond the fortified Federal posts. Hurst treated Green with courtesy during the encounter, even sharing his canteen of whiskey with him. Just before he was seized, Green hid his pistol in a mattress in one of the beds at the Barret Farm. Rebecca Barret came across the weapon and gave it to Green's family after Hurst's men had departed. When Green returned home months later, after escaping from a prisoner-of-war camp, his pistol was returned. Following the war, the Barret family generally prospered. They added acreage to the farm and built a solid foundation for the family's diverse twentieth-century enterprises which included the store (in operation until the 1980s), a cotton gin and delinter, and the Barretville Bank & Trust Company.

From *The Soldier in Our Civil War*

END NOTES

Introduction

Ash, Stephen V. *Middle Tennessee Society Transformed 1860-1870, War and Peace in the Upper South* (Baton Rouge: Louisiana State University, 1988), 13-14.

Still Hollow Farm

West, Carroll Van, and Elizabeth Moore. "Allen-Birdwell Farm, Greene County, Tennessee." National Register of Historic Places Nomination Form. Tennessee Historical Commission. Listed 2011.

Fermanagh-Ross Farm

Brock, Daniel, and Robbie D. Jones. "Maden Hall Farm, Greene County, Tennessee." National Register of Historic Places Nomination Form. Tennessee Historical Commission. Listed 2009.

Tarwater Farm

Tarwater, Charles B. *A Brief History of the Tarwater Family* (Knoxville, 1957).

Fowler-Lenoir Farm

McBride, Robert M. *Biographical Dictionary of the Tennessee General Assembly, Vol. II* (Nashville: TSLA & THC, 1979), 304.

Barret Farm

McCalla, Jon P., ed. *An Illustrated History of the People and Towns of Northeast Shelby County and South Central Tipton County: Salem, Portersville, Idaville, Kerrville, Armourtown, Bethel, Tipton, Mudville, Macedonia, Gratitude, Barretville, and Rosemark, Tennessee* (Memphis: Historic Archives of Rosemark and Environs, Inc., 2010), 317.

Refer to the "Bibliographical Essay" for additional resource materials.

The Farm As Battleground

The landscape of Tennessee was a battleground throughout the war as hundreds of military engagements took place in all sections. Fort Donelson, Shiloh, Stones River, Chattanooga, Knoxville, Franklin, and Nashville were major battles, while lesser-known but still deadly skirmishes occurred almost everywhere. The total recorded number of confrontations in the state exceeds those in all others except Virginia, giving Tennessee the most fields of battle in the war's Western Theatre.

Mounted troops, infantry, and artillery were dangers to anyone and anything in their path, but to those families on whose farms opposing forces met, the war entered their homes, their farm buildings, and their fields, leaving behind destruction and death. For many slaveholders, the arrival of Union forces also meant that the farm's slaves escaped in scores to new lives behind Federal lines. With thousands of Union and Confederate soldiers travelling east, west, north, and south for four long years, farm families were often in the direct line of fire with little protection, mercy, or aid during the clashes or afterwards.

The immediate consequences of combat were particularly horrible to the families who saw the dead, dying, and wounded all around their farms and in the rooms of their homes. Women and children, who normally went about their daily lives in these places, tended the sick and the dying as best they could. Years later, individual recollections of farms as battlegrounds, hospitals, and cemeteries were among the most riveting of stories passed along to children and grandchildren.

Farms became battlegrounds and were overrun by soldiers who left desolation and death in their wake.
From *Leslie's*

Elmwood Plantation
Rutherford County

Not far from the geographic center of Tennessee, Thomas Hord, an attorney from Hawkins County, purchased an 840-acre Rutherford County plantation in 1842. Situated on gently rolling hills and fertile bottomlands along Overall Creek, the Hord Farm soon became a substantial agricultural producer. The farm stood on both sides of the Nashville-Murfreesboro Turnpike, and the Nashville & Chattanooga Railroad was built through the property in 1851. This excellent access to transportation further boosted the farm's production in the 1850s.

Hord married twice and had a total of eleven offspring. His first wife, Mary McCulloch, died in 1851 after bearing nine children. Eight years later, Amelia Mildred Gilmer became the widower's second wife, and two more children were born into the family. The Hords lived in Elmwood, a substantial Greek Revival-style brick house that occupied a slight knoll dotted with mature trees. By 1862, the Hords were producing sizeable cotton, corn, wheat, and hay crops, and raising herds of livestock along with poultry. The plantation relied on slave labor which by 1860 included sixteen children under the age of ten along with twenty-nine males and females from age ten to sixty-two. Hord, a Unionist, noted in his petition for redress following the war that in August of 1862, his "men were taken to work on the stockade forts here and in Murfreesboro until the army fell back from Alabama to Kentucky." He claimed pay for their service and for those who did not return.

Civil War military strategy often focused on the protection of the railroads to ensure sufficient logistical support for large armies in the field. The Union occupation of Nashville in 1862 and the presence of Gen. Braxton Bragg's Confederate army near Murfreesboro meant that Elmwood lay directly in the path of converging forces and would pay the price. The opposing armies met in two major battles on the Hord plantation, first in 1862 and again in 1864.

On Monday, December 29, 1862, the Union army marched by the Hord farm to confront Bragg's troops. Two days later, during the Battle of Stones River, skirmishing occurred in the yard of the main house. During the fighting, the Confederates briefly occupied the Hord house until they were dislodged by the 3rd Ohio Cavalry. Hord wrote that during the week he furnished twenty-one horses and mules, thirty-seven head of cattle, one hundred hogs, nine hundred barrels of corn, and at least forty tons of hay and fodder. By Saturday, January 3, 1863, the farm was left with little food to last the rest of the winter.

The location (see star) of Elmwood Farm on either side of the Nashville Turnpike and on the Nashville & Chattanooga Railroad placed the Hord family and their farm in jeopardy on many occasions, but squarely in the thick of battles in late 1862 into early 1863 and in 1864.
From *The Soldier in Our Civil War*

Elmwood sheltered the Hord family during two battles and was used as a field hospital.

Elmwood served as one of the Union army's field hospitals. Thomas Hord stated in his postwar damage claim that over 6,000 wounded soldiers passed through his farm. The family was relegated to one room while amputations were carried out in the back parlor by regimental surgeons. Several Union physicians and many more wounded patients stayed on at the Hord farm until March 1863.

During the Battle of the Cedars on December 4, 1864, part of Confederate Gen. John Bell Hood's overall advance on Nashville, the Hords were again in the thick of the fighting. This attack happened in the fields between the mansion house and Overall Creek. Confederate forces under Gen. William Bate of Nathan Bedford Forrest's command attacked Federal Blockhouse #7 protecting the railroad bridge over Overall Creek. The 13th Indiana Cavalry, under Col. G. M. L. Johnson, set up a skirmish line on the south bank of Overall Creek to observe and impede the advance of the Confederates. Maj. Gen. R. H. Milroy was dispatched from Fortress Rosecrans in Murfreesboro to bolster Johnson's cavalry and the ensuing fighting east of the Hord house and all along the creek was fierce. Both sides suffered severe losses. Ultimately, on Dec. 7, Gen. Forrest passed down the Nashville Pike to attempt a raid on Murfreesboro, and Gen. Milroy broke the rebel line, forcing Forrest to retire from the field. Blockhouse #7 held out for two weeks, and on December 15, Hood retreated from Nashville to end his Tennessee campaign.

At Elmwood, the military engagements and regular foraging expeditions devastated the farm's fields, forests, and buildings, as well as the family's mental and physical condition. Troops on the march took corn, cotton, wheat, hay, fruit, and considerable numbers of livestock. They also destroyed two large barns, a dwelling house, three slave houses, six log stables, and nine other substantial outbuildings. The war also overturned the traditional system of slave labor that southern agriculturists had used to produce such wealth. Although Hord submitted claims to the United States government for his losses, the family received only a small amount in 1911.

Thomas Hord died in 1865, and his estate was divided equally between his wife, Amelia, and eight surviving children. She ran the farming business and sold off land in Louisiana and Arkansas to keep the Tennessee operation financially afloat in the years following the war. The Hords planted food crops, continued raising and selling horses and cattle, and gradually increased their commercial cotton production. The fields and pastures of the farm, once overrun by soldiers, are now worked by the fourth and fifth generations of the family. The Hord Farm, listed in the National Register of Historic Places, is located near Stones River National Battlefield.

The Battle of Stones River covered the Hord Farm
From *Leslie's*

Glen Leven
Davidson County

Glen Leven, located just four miles south of Nashville, was a sprawling plantation of nearly 1000 acres owned by John Thompson. A man of vision, he understood the importance of transportation, economics, and diversity in crops and livestock to agriculture. He supported and invested in the Franklin Turnpike Company, created in 1829, and his property was positioned on this first major turnpike in Middle Tennessee when it opened in 1838. Not content with marketing by road, he was an initial member of the board of commissioners for the Nashville & Alabama Railroad and also for the Mississippi & Nashville Railroad. Glen Leven was situated between the turnpike and the railroads, providing an enviable location for shipping, receiving, and travel.

Thompson was within the less than five percent of Tennessee farmers who could be classified in the planter class, and in 1860 owned sixty-two slaves. The Tennessee Agricultural Census of Davidson County of that year listed Thompson with 430 acres of improved land, 445 acres of unimproved land, and $4,650 worth of livestock. His land holdings were valued at $105,000, and his personal property was reported as $263,050. He depended on slaves to work the farm and tend to the household and gardens. John and his wife, Mary, and their children lived in an impressive Greek Revival house that was completed about 1857. With the onset of the Civil War, the antebellum social structure as well as the daily life and fortune of the Thompsons changed quickly as Nashville came under Federal control in February of 1862 and remained so for the duration of the war.

A strong Confederate sympathizer, Thompson was too old to serve in the army and his sons were too young, so he made substantial contributions of money and equipment. For his allegiance, he was fined and imprisoned by Union authorities in Nashville, most likely in April of 1863, and remained there until the end of the war. Mary Thompson and the children, ages ten and eight, remained on the farm where she astutely managed the family and property.

Glen Leven continued to be an active place during the occupation, and Mary received callers, hosted gatherings, and also boarded Union soldiers, including Gen. Gates P. Thruston, who met his future wife, Ida Hamilton, at one of these social events. Though she initially snubbed the officer, who would become a distinguished archaeologist and author, an eyewitness noted that "he retaliated by marrying her."

Foragers, primarily occupying Union forces, focused on Glen Leven's fertile fields, well-stocked barns,

Built by John and Mary Thompson in 1857, Glen Leven house and farm were in the thick of fighting during the Battle of Nashville.

John Thompson, photographed about 1860, was imprisoned for most of the Civil War.

and large livestock herds time and again. It was, however, the Battle of Nashville in December of 1864 that brought fierce fighting and all of its calamities to the farm and family. As the armies under the command of Confederate Gen. John Bell Hood, moving north from the devastating Battle of Franklin, and Union Gen. George Thomas converged on the state capital, soldiers were encamped all around the house and in the fields of Glen Leven. A neighbor, Mary Vaulx, who lived north of Glen Leven, recalled that "it seemed as if a thousand fires were going in John Thompson's Woods."

As the battle commenced and the troops met and fought with increasing desperation in the bitter cold mid-December weather, Mary and her sons remained in the house. She was, however, often seen on the battlefield tending to the wounded, and she had the foresight to prepare several rooms in Glen Leven to receive the injured. The back parlor became an operating room, and Mary was asked to help with the amputations; the surgeons reportedly even used "the piano as an operating table."

The dead from both sides, including at least thirty-two soldiers of the United States Colored Troops, were hastily interred on the grounds and in the fields of Glen Leven. As the remaining troops moved on, Glen Leven was a stained and scarred piece of land that served as a cemetery for men who perished in the Battle of Nashville [later they were removed and reburied at Woodlawn Cemetery and the Nashville National Cemetery].

When John Thompson was released from prison at the end of the war, he finally saw firsthand the nearly complete physical destruction of his farm. He also was billed for taxes on his real estate during the war years and an additional tax to aid Union widows and orphans. Both John and Mary Thompson had kept careful records of what had been taken from their property, asking for signed receipts from officers who led foraging parties. When Thompson was given the bill for his taxes, he presented the numerous receipts for produce and livestock that had been taken from his farm, and they were accepted in lieu of cash payments for back taxes. Because of the Thompsons' foresight in requesting and saving these documents, Glen Leven was likely saved from immediate financial ruin. Thompson also filed claims with the federal government after the war but had no success.

John Thompson was able to recoup much of his fortune and revitalized his land in the decade following the war. For example, his investments in railroads, which were quickly being repaired and expanded, proved to be sound, and he began rebuilding the farm. At his death in 1876, he left a valuable estate. John M. Thompson, Jr., who had remained with his mother at Glen Leven during the war years and came of age as his father was returning the farm to production, was also a progressive and successful farmer who served in the Tennessee legislature from 1893 to 1903. Glen Leven remained in the Thompson family until it was willed by Susan Thompson West to the Land Trust for Tennessee in 2006. With this arrangement, the agricultural future of this historic farm is secure without the specter of commercial and industrial development.

Jacobs Farm
Coffee County

Farmers in southern Middle Tennessee also experienced the trials of combat on their property during the Tullahoma Campaign in 1863. Northeast of Beech Grove, along Garrison Fork, Jeremiah and Rebecca Rudd Jacobs began farming about 900 acres in the 1820s. The land supported livestock – cattle, horses, and sheep – as well as small grains and some cotton.

In 1855, after the death of the elder Jacobs, their son, Alfred, became the sole owner of the property. Alfred Jacobs, who married Catherine Dillard in 1842, owned and operated Jacobs Store for thirty years. The establishment was well known and attracted customers for miles around. Like many good businessmen, Jacobs tried to remain neutral when hostilities broke out and to sell to whomever was buying. The store's location on the Manchester Pike and Murfreesboro Pike at the intersection of the Wartrace Road on the southern side of Hoover's Gap placed it in the thick of the fighting during the Battle of Hoover's Gap on June 24-26, 1863. At various times, officers from both armies requisitioned the residence for their quarters while their regiments camped in the nearby fields and woods. After the soldiers arrived, fence rails and wandering livestock were the first things to disappear.

When Col. John T. Wilder's "Lightning Brigade" of U.S. Gen. George H. Thomas's corps burst through the gap on June 24, they surprised C.S. Gen. Alexander P. Stewart's division of the Army of Tennessee that had been encamped along Garrison Fork. Within two days, Thomas's advance had driven Stewart's forces out of the area completely. The retreating Confederates marched past Jacob's Store on their way to Fairfield. Despite their proximity to the severest fighting of the campaign, Jacobs's buildings suffered little damage. Alfred

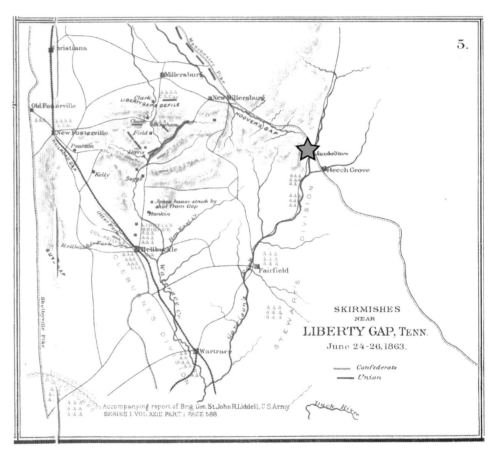

*The skirmishes at Liberty Gap (note the location of Jacob's Store marked by star) also spilled onto the **Knight Farm** in **Bedford County** in June 1863. Robert Walton Beachboard, who had purchased the farm just six months before, and his wife, Elizabeth, moved farther into Hatchett Hollow to live with neighbors when hostilities seemed imminent. When the family returned to their home after the fighting ceased, they discovered fresh graves in the family cemetery. From The Official Military Atlas of the Civil War*

continued to operate the store and was on the committee to rebuild the Coffee County courthouse in 1870. Eventually the farm passed to Alfred and Catherine's children and to their descendants who own and operate the farm today.

Samuel Raulston Farm
Marion County

The southeastern part of the state was also the scene of terrible battles in and around the strategic transportation junction of Chattanooga. About seven miles northwest of what is now South Pittsburg is Sweeten's Cove (also known as Sweden's Cove). Here Col. James Raulston (Roulston), an early settler of Marion County and a veteran of the Creek Wars as well as the War of 1812, and several of his sons received land grants of several thousand acres at the southern end of the Sequatchie Valley. Adjacent to Raulston's property were large tracts of land owned by another early settler, Capt. Robert Beene (Bean). Raulston and his family raised corn, wheat, oats, cotton, timber, cattle, hogs, sheep, horses, and mules. Samuel Raulston, one of fourteen children of James Raulston, married Millie Beene in 1835, and they were living on their own farm, about eight miles from that of his father, by 1841. Though slaveholders, apparently the large family of Samuel and Millie, which eventually included ten children, farmed without the help of adult slave labor. In 1850, Samuel Raulston is listed with two slaves, a female aged 14 and a one-year-old male, and in 1860 he owned one female, aged ten.

On this 1863 map, the farms of the related Beenes and Raulstons (Bean and Ralston) are noted in Sweeten's (Sweeden's) Cove along the Jasper-Sewanee Road through the Sequatchie Valley (see star). Though the families differed during the Civil War, they put hard feelings behind them and pulled together after hostilities ended.
From *The Official Military Atlas of the Civil War*

As the war came to the valley, strained relations existed between the two neighboring families. Five of the Raulstons, who remained loyal to the Union, had married five of the Beenes, who supported the Confederacy. Two of the Raulston boys, William Henry and Sam Houston, enlisted in Col. William B. Stokes's 5th Tennessee Cavalry (USA). William Henry was killed in December 1862, a few days before the Battle of Stones River, and is buried in the National Cemetery in Nashville. Sam Houston Raulston survived the war and returned to the farm.

On June 4, 1862, a small skirmish took place in Sweeten's Cove between the troops of U.S. Gen. James Negley and the cavalry of C.S. Col. John Adams. In the Bean-Raulston Cemetery, within a mile of the circa-1850 Primitive Baptist Church of Sweeten's Cove, is a memorial marker to twenty unknown Confederate soldiers who died in the engagement. The fighting took place, in part, on the Raulston Farm.

Family history recounts that the Raulstons "saw thousands of troops march by their house by way of the Jasper-Sewanee Road on their way to battles at Chattanooga and Chickamauga." Both "Union and Confederate troops ravaged the land, taking all of the livestock and crops they could." Correspondence between Lt. James B. Roulston, likely a relation of Samuel's, and Lt. Col. Robert Galbraith of the 1st Tennessee Cavalry (USA) gives some insight into the situation of farm families in Marion County. In a communication from August 1863, Roulston asks for permission for himself and 30 men of his company to visit and provide for their families in Marion County.

Bean-Raulston Cemetery

Some, including Raulston, had lost their homes and much of their property to marauding Confederate soldiers. He closed his appeal by saying that this visit would enable the men to relieve the needs of many suffering women and children. His request was granted.

Samuel Raulston died in 1866, and several stories have been passed down about how his widow and children tried make a living in the hard times following the war. To survive, they travelled as far as McMinnville to barter for material for clothing. Salt was in such short supply that they, like other farm families, sifted the dirt from the floor of the smokehouse and boiled it to obtain salt that had dripped down and accumulated there over the years. The fifth generation of Raulstons now manages the farm that endured conflict both within their extended family and with the armies that marched through and fought in Sweeten's Cove.

The Bean-Raulston Cemetery holds the remains of Confederate soldiers killed in 1862 on and near the Samuel Raulston Farm.

SACRED GROUND

For most communities, the cemetery is the oldest evidence of settlement. The same is true of farms that have family cemeteries. Here are buried farm founders, generations of their descendants, members of the extended family, and neighbors. If slaves were held on the property, their graves may be nearby, though are rarely marked with more than a fieldstone.

During the Civil War, soldiers' remains were returned to the family, if possible. National cemeteries include the graves of Union soldiers, while Confederate casualties were sometimes taken from hasty burials on the battlefield and reinterred in sections of larger cemeteries near the town where they died. Individual markers for soldiers and veterans of both armies, including United States Colored Troops, can be in found in cemeteries, large and small, across Tennessee. As freed men and women established their own communities after emancipation, their burying grounds were also first located on their land. Cemeteries are among the most enduring historic sites on the landscape. Many burial grounds have been recorded and photographed to assist genealogists, historians, and family members searching for ancestors and their stories. Whether long abandoned or carefully maintained, each of these places is sacred ground.

*In **Overton County**, the cemetery at the **Looper-Thompson** Farm contains rare "comb" markers that are unique to the area. Buried here is Joseph Looper whose family of French Protestants (Huguenots) fled their native country in search of religious freedom. Joseph purchased property in Overton County that he had discovered on previous hunting trips to the area. The question of slavery was a difficult one for the Looper family to reconcile. After his death in 1858, Joseph's will provided for two slaves, James and Julia, and their three children to be set free, and be conveyed to a port where they could get safe voyage to Liberia "or some of the African regions that will receive them as free people." His will was contested, however, and the case eventually reached the Supreme Court. The slaves were freed and are listed as such on the 1860 census, but apparently never left for Africa. The Looper farm where William, a son of Joseph, and his family lived was raided many times during the Civil War, and the family learned to hide food and supplies in a nearby cave. On one occasion, raiders ordered William to give them all of his money or be killed. William wanted to resist, but his wife, who had hidden the money, complied and the raiders left without harming the couple.*

*The marker for John Alphonso Walker is typical of Civil War headstones that note the rank and service as well as the usual birth and death dates. Walker and his wife, Ann Willey, and their three children migrated from Liverpool, England in 1848 by way of New Orleans. They eventually settled in Nashville where Walker, trained in making all types of leather articles by hand, was the foreman of Morrow Brothers saddle shop. During the Civil War, the shop made saddles and bridles for the Confederate cavalry. In 1869, Walker purchased a farm of nearly 400 acres in **White County** in Yankeetown, so called because Federal troops camped there during the war. He named his land **Dixie Farm**. Walker's great-great grandson owns some of the original acreage today, including the cemetery where the veteran is buried.*

The Hoskins family began farming on the Old Port Royal Road in **Montgomery County** in 1819. Neander Hoskins shared in the inheritance of the farm with his siblings, receiving the homeplace when his father died in 1840. He and his wife, Margaret, had 10 children. Son John Gaston Hoskins, born in 1841, enlisted in November 1861 at Clarksville with the Tennessee 49th Infantry (CSA) and was captured at Fort Donelson in 1862. He was sent to Camp Douglas, Illinois, and was exchanged at Vicksburg, Mississippi. He remained in the hospital there and was discharged in December with the comment, "He will never be able to walk and will be a cripple for life." The farm cemetery contains the graves of slaves and family members, including John Gaston Hoskins, who lived with his infirmities until 1896. Descendants own and operate the land as the ***Anderson Farm***.

H.E.F. Blair and Sam Blair Farms
Loudon County

Control of the Tennessee River, particularly at ferry crossings and railroad bridges, was a focus of the armies in East Tennessee. Blair's Ferry, established by James Blair by 1815, operated on both sides of the Tennessee River and provided a way into lands occupied by the Cherokee Nation. Blair's landholdings were extensive, and at least five certified Century Farms evolved from his properties. When Blair died, his son, Wiley, inherited a considerable amount of property in both the town of Loudon and along the river.

As the younger Blair reached maturity and prospered during the early 1850s, he and his wife, Mary Johnston, whom he had married in 1843, began building an impressive country home facing the river. The contractor, Alexander McInturff, used slave labor to make the bricks for the foundation and chimneys. Blair died of cholera in 1854, but his widow oversaw the completion of the spacious house by 1857. The farm's strategic location overlooking the Tennessee River, with both a ferry landing and the East Tennessee & Georgia Railroad bridge, ensured that the Blairs would be severely affected by the war. As the war came to East Tennessee, Mary Blair and her children, though far better off financially than many other families, faced challenges and losses that could not be calculated in currency.

The Blair house and farm was the scene of military activity throughout the war because of its location near the Tennessee River. The house originally faced the river, but was remodeled in 1935 by Laura Blair Vance, Kate Blair, and Albert Blair to front the Lee Highway.

Among the Blair papers is a poignant letter to the family dated August 25, 1862, from Lt. Francis C. Shropshire, adjutant to Lieut. Col. John S. Fain of Smith's Legion, Georgia Partisan Rangers. The missive addressed the death of Rachel, the Blairs' thirteen-year-old daughter, who "died of disease contracted while engaged in missions of love and mercy, to the sick and dying soldiers in our hospitals." Shropshire mentioned that representatives of the battalion would attend the funeral in respect to her memory.

In 1863, Mary Blair and her remaining children were forced to move to their townhouse in Loudon while Union officers used the farmhouse as a headquarters. While the Federal soldiers camped on the farm, they constructed four redoubts and a stockade – with trenches and cannon emplacements – to protect the 1,670-foot railroad bridge that spanned the river. Several cavalry skirmishes took place on the farm as the opposing forces vied for control of this strategic position, and the house was used as a hospital on several occasions.

After the war, the Blair family returned to a changed and devastated landscape. For example, the grove of trees that formerly had extended from the porch to the river, a distance of more than a mile, had been cut down completely by Union infantrymen. After the war, the Blairs submitted unsuccessful claims to the government for the loss of 40,000 fence rails and 125 acres of timber taken from the farm for firewood and for building "no less than 250 winter huts for federal troops during the winter of 1863-64." In the twentieth century, this farm was divided between heirs, and the two farms are owned and operated by Blair cousins. Over the years, the family has collected material evidence from the period when their farm was a place of combat.

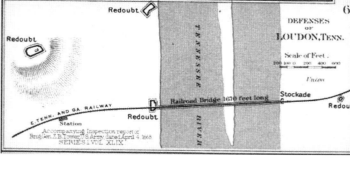

From *The Official Atlas of the Civil War*

Within the grove of trees on the Blair Farms is the remnant of one of four redoubts built to protect the East Tennessee & Georgia Railroad bridge that spanned the Tennessee River.

McQueen Farm
Loudon County

Thomas Jefferson and Eliza Kerr Mason established a farm of 292 acres along the Tennessee River in 1852, just outside of Loudon which was at that time was in Roane County. As a young man, Mason was in the Georgia Militia and then in the United States Army where he commanded a detachment of soldiers who rode with the Cherokee people on the forced "Trail of Tears" march to Oklahoma in the 1830s. Returning to Tennessee, Mason married Eliza Kerr, daughter of an Irish immigrant, in 1845, and operated a flatboat on the Tennessee River. The Mason property was subjected to an unusual amount of military activity because of its location and considerable river frontage. Huff's Ferry docked at what is now the McQueen Farm when it made its regular river crossing transporting people, wagons, and supplies to the Loudon side. In November 1863, 15,000 infantry and artillerymen, under the command of Lieut. Gen. James Longstreet, landed on the Mason Farm after crossing the river on a pontoon bridge.

Mary Jane Mason, like her neighbor, Rachel Blair (see previous page), was in her teens when she nursed sick, wounded, and dying soldiers. After the war, Mary married Alexander M. Presnell in May of 1874; he died less than a year later of typhoid. Mary graduated from Grant Memorial University in Athens, Tennessee, then taught at Loudon High School before becoming the Chair of English Literature and History at Chattanooga University. Mary Jane Mason Presnell and her brother, Thomas Jefferson Mason, Jr., died of tuberculosis just 11 days apart in late 1888 and early 1889. Their sister, Elizabeth "Bettie" Mason, married Edmund Preston McQueen in 1882, and portions of the original Mason farm, including the site of the ferry crossing, remain within that family today.

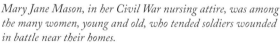
Mary Jane Mason, in her Civil War nursing attire, was among the many women, young and old, who tended soldiers wounded in battle near their homes.

The Mason Farm received thousands of troops in 1863.

Massengill Farm
Grainger County

Moving further to the northeast, the Massengill Farm dates from 1796, the year Tennessee became a state. Michael Massengill was both a farmer and a miller who constructed a grist mill that served the Buffalo Springs community southwest of Rutledge. A slaveholder, Massengill differed with many of his neighbors on the question of secession; Grainger County residents were mainly pro-Union. Soldiers from both sides claimed this industrial site with its tall building that afforded views in all directions and that offered a good supply of water for men and horses. Union officers occupied the mill house as headquarters during the Battle of Bean Station on December 12, 1863. A few days later, the Army of Tennessee's 11th Cavalry, moving through Bean's Station and Bull's Gap seeking out Federal forces, reported that on the sixteenth, it "captured 16 prisoners at Massengill's Mills in Grainger County." Union sympathizers torched the farm house at the close of the war, but the mill survived and continued to be operated by the Massengill family until the 1940s.

Massengill's Mill, occupied at various times by both Union and Confederate troops, survived the war.
Courtesy *Grainger County Archives*

Fairview Farm
Jefferson County

The Battle of Mossy Creek Marker

The Battle of Mossy Creek took place on and around Fairview Farm. Nearby is a Tennessee Historical marker, and the Branner Cemetery where Capt. E. J Cannon (USA), who was killed during the fighting, is buried.

The battle that occurred at Mossy Creek, the original name of Jefferson City, on December 29, 1863, was fought, in part, on and around Fairview Farm. The fighting resulted from a Federal advance of more than six thousand soldiers from Strawberry Plains on December 18, 1863, to pressure the Confederate army of Lt. Gen. James Longstreet, headquartered for the winter in Russellville (Hamblen County). Longstreet's cavalry, commanded by Gen. William Martin, patrolled a twenty-five-mile arc that ran from Rutledge to Dandridge and centered on Mossy Creek.

The residence at Fairview was a well-designed and substantial brick house built about 1850 by Stokley Donaldson Williams and his wife, Mary Porter Reese. At the time the house was built, the Williams owned considerable acreage that was farmed with the help of slave labor. In 1850, six slaves, three males and three females ranging in age from one to forty-five, are listed. Eleven slaves are recorded in 1860; four were under the age of ten and the oldest male was forty. The residence had extensive gardens and a boxwood-lined drive led from the house to the road. These plantings were all destroyed during the war. At various times during the years of conflict, the family had to vacate their house because of fighting on and around their farm; the Battle of Mossy Creek, however, was the fiercest the residents of Fairview experienced.

A history of the 1st Regiment of the Tennessee Volunteer Cavalry (USA) describes the scene in which soldiers "fell back to the residence of Stokely Williams, a large, two-story, brick house, and during the engagement it was struck several times by flying shells." The fighting left the Union with at least 165 casualties, and the Confederate dead numbered about 400. Even today, Fairview carries bullet scars from the fighting. The historic farm and house has been in the ownership of the current family since 1894 when Lissie Reed James purchased the 370 acres and the house from the Williams family. The house is listed in the National Register of Historic Places.

The Williams home at Fairview was the scene of fierce fighting during the Battle of Mossy Creek.

Moore Farm
Hawkins County

The East Tennessee & Virginia Railroad, built largely by slaves, linked Knoxville and Bristol by the late 1850s. At Bull's Gap, the line ran within a few miles of the Moore Farm. John Rufus Moore was born in 1843 on the farm his parents had established in 1835. In June of 1864, John enlisted in the 3rd Tennessee Mounted Infantry (USA) and was posted as a guard along the railroad. The control of this important rail line was one of the contributing factors to the Battle of Bull's Gap in November 1864, when fighting occurred around and on the Moore farm.

This engagement was part of Confederate Maj. Gen. John Breckenridge's campaign to drive Union forces from that part of the state. At the same time, Union Brig. Gen. Alvan Gillem planned to rid the area of Confederate forces. Their troops collided at Bull's Gap. It was along the railroad and the surrounding countryside that much of fighting took place on November 11 and 12. Descendant, Howard G. Moore wrote in 2003 that "one can still clearly see a road that was made by soldiers during the Civil War around a bluff on the farm." He also advised that family papers were "destroyed by the soldiers" during this time.

In 1866, John Rufus Moore built a house on the farm, and then, in 1871, he married Emily McCullough, also of Hawkins County. Moore farmed after the war but also had a lucrative business selling agricultural implements to a community that was rebuilding. The East Tennessee & Virginia had consolidated with the East Tennessee & Georgia Railroad in 1869, giving travel and trade a broader base in the area. Moore ordered and then received shipments of parts which were delivered to Rogersville Junction near his farm. He then transported them the short distance to his shop, where he assembled the tools and equipment before selling them to local farmers. The Moore house and portions of the working farm are listed in the National Register of Historic Places.

In 1866, just two years after soldiers had fought here, John Rufus Moore built this two-story frame house with a rear ell, along with an adjacent well and smokehouse. The postwar buildings remain on the farm owned and operated by his descendants.

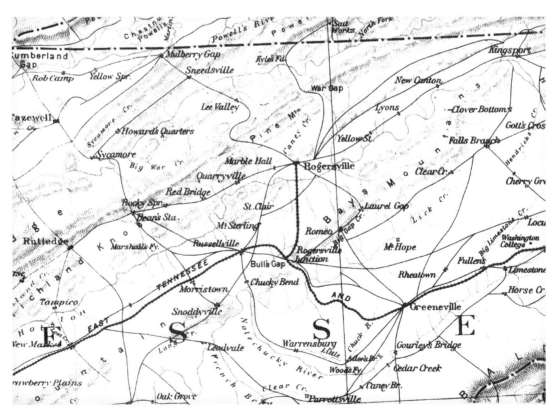

The Moore Farm, located near the Rogersville Junction of the East Tennessee & Virginia Railroad at Bull's Gap, was the scene of fighting in 1864.
From *The Official Military Atlas of the Civil War*

END NOTES

Elmwood Farm

King, Spurgeon. "Elmwood (boundary increase and additional documentation) Rutherford County, Tennessee." National Register of Historic Places Nomination Form, Tennessee Historical Commission. Listed 2008.

Hughes, Mary B. *Hearthstones: The Story of Historic Rutherford County Homes.* (Murfreesboro, Mid-South Publishing Co. Inc, 1942), 49.

Glen Leven

Tennessee Civil War National Heritage Area. *Glen Leven: Historic Structure Report and Archaeological Reconnaissance Survey, Center for Historic Preservation, Middle Tennessee State University,* 2011, 15-31.

Massengill Farm

Fisher, John W. *They Rode with Forrest and Wheeler: A Chronicle of Five Tennessee Brothers' Service in the Confederate Western Cavalry* (n. p., 1995), 68.

Fairview

Carter, V. R. *History of the First Regiment of Tennessee Volunteer Cavalry in the Great War of the Rebellion, with the Armies of the Ohio and Cumberland, Under Generals Morgan, Rosecrans, Thomas, Stanley, and Wilson, 1862-1865* (Knoxville: Gaut-Ogden Co., Printers and Binders, 1902), 128.

The Garden Study Club of Nashville, *History of Homes and Gardens of Tennessee* (Nashville: Parthenon Press, 1936; Reprint Edition, 1964 by Friends of Cheekwood), 64.

Moore Farm

Sharp, Leslie N. "Moore Family Farm, Hawkins County, Tennessee." National Register of Historic Places Nomination Form, Tennessee Historical Commission. Listed 2006.

AmericanCivilWar.com. "Battle of Bulls' Gap, Tennessee." http://americancivilwar.com/statepic/tn/tn033.html.

Refer to the "Bibliographical Essay" for additional resource materials.

The Hard Hand Of War

Conflicting loyalties often lead to heated words of disagreement and, at the worst, to acts of violence resulting in the destruction of property and death. When the United States plunged into civil war, conflict enveloped Tennessee's farm families in their own households and with relatives, neighbors, Union and Confederate troops, renegades, bushwhackers, and guerillas. Rivalries, betrayals, and savage incidents occurred across the state and resulted in consequences that survive today in local and family history.

In the course of any war, some occurrences can only be described as atrocities. The American Civil War had a full measure of horrific actions that affected civilians and soldiers on both sides. Some of these stories have been well documented, while other accounts come from long-standing oral traditions within families. Whatever the reasons or circumstances, some farms were the scenes of particularly cruel acts – deeds that brought the full horror of war crashing down on whomever happened to be in that place.

For many Tennessee families, the terrors and direct effects of the conflict did not end with Appomattox but continued to influence the operation and lineage of farms for generations. That these incidents and events are remembered more than 150 years after they happened is a testament to the incalculable cost of war.

(Opposite page) From *The Soldier in Our Civil War*

Officer Farm
Overton County

William A. Officer was a successful farmer and stock man, and a slaveholder, on the Upper Cumberland Plateau. Unlike most other agricultural holdings in the area that tended to be small, subsistence operations, the 1,000-acre Officer Farm was one of the largest in the county. In 1850, Officer owned livestock valued at nearly $2500, and raised over 2500 bushels of corn along with oats, wheat, sweet potatoes, and flax. Seven slaves, ranging in age from two to thirty years, lived on the farm. Officer, an active supporter of the Confederate cause, had a reputation for harboring soldiers, and the fact that he continued to do so after Union occupation in the region made him a prime candidate for punishment. Federal forces periodically seized private property and burned homes of secessionists known or suspected to be in complicity with guerrilla bands. On February 28, 1864, Federal soldiers compelled Officer to take the oath of allegiance to the United States at Sparta. Harsh consequences sometimes awaited those citizens caught aiding the Confederacy after taking the oath, and the Officer Farm soon found itself the target of Union cavalry determined to curtail assistance to the enemy.

In June 1862, former DeKalb County Congressman William B. Stokes had been authorized to raise the 5th Tennessee Cavalry regiment as part of the Union army's effort to counter the increasingly widespread Confederate guerrilla activity in the Upper Cumberland. The local men who made up the unit were quite familiar with their

The Officer Farmhouse and Cemetery, with six gravestones from 1864, are reminders of the violence that took place here and in other counties along the Cumberland Plateau.

Confederate neighbors. On March 12, 1864, Stokes sent Capts. Joseph H. Blackburn and Shelah Waters with 200 men on a scouting trip up the Calfkiller River Valley in search of the enemy.

Officer's wife and daughter had risen early that morning to prepare breakfast for seven Texas Rangers led by Lieut. Robert S. Davis. These men had been cut off from their regiment during earlier fighting and remained in the locality, harassing residents who remained loyal to the Union. They were preparing to eat when the Federal troopers rode up, entered the house, and began shooting. Stokes sanctioned the killings, noting that the rangers were "of the most daring and desperate character." Stokes continued, "These men had been murdering and robbing Union citizens" in the region. After killing the soldiers, the Union horsemen attempted to burn the house, but Officer extinguished the fire before it could spread.

Also present that day, but hiding, was the Officers' son, John, who served in Col. James W. Gillespie's 43rd Tennessee Mounted Infantry. The younger Officer, who had been paroled at Vicksburg, avoided detection and eventually returned to his regiment. After Union forces departed, Abraham H. Officer, a slave who witnessed the incident and later provided details of what had transpired, loaded the dead soldiers onto an ox-cart and carried them to the family cemetery for burial.

In the period following the war, William Officer had little left except his land but quickly regained his prominence in the commercial world as a farmer and livestock trader. Working with Officer was Abraham and another former slave, Robert, who continued to live on the farm into the 1870s. At the time of his death in 1886, Officer remained one of the most prominent landowners and farmers in the Upper Cumberland.

Bell Farm
Unicoi County

The pendulum of violence could and did swing both ways. Dr. David H. Bell, a native of Ireland, and his wife, Sarah McKeldin, established a substantial farm of some 2,000 acres in the Limestone Cove community of what was then Carter County (Unicoi County would be formed in 1875). The Bells raised wheat and different types of livestock on their plantation. Their four children, as well as David's bachelor brother, James, lived and worked there.

One November morning in 1863, the Bell family was preparing to feed a group of about fifty unarmed North Carolina civilian refugees who were heading toward Kentucky to join Federal forces. Suddenly, Confederate conscription officer Col. V. A. Witcher, Jr., with soldiers of the 34th Virginia Cavalry, surrounded the farm and gunned down at least eight of the travelers. The slain were B. Blackburn, Calvin Cantrel, Elijah Gentry, Jacob Lyons, Wiley Royal, John Sparks, and two who were never identified. James Bell was dragged from the house and, according to the family and the authors of *The History of the 13th Regiment, Tennessee Volunteer Cavalry, U. S. A.*, was shot and "after wounding him his head was laid on a stone and his brains beaten out until they bespattered the ground all about his body."

Mrs. Bell was pistol-whipped, and the family's large brick house was ordered "to be set on fire which was done."

A plot of land across the road from the burned house became the burying ground for the slain. The eight men were buried in a mass grave while James was interred in a separate plot. David and Sarah Bell weathered the war years and kept their land intact. The Agricultural Census of 1870 lists the total worth of the Bell farm at $4000. They owned four horses, four milk cows, a flock of eighteen sheep, and ten hogs. They produced 120 bushels of wheat and 500 bushels of corn in that year as well as vegetables and fruits.

Buried in the same cemetery with the slain men are David Bell, who died in 1892, Sarah Bell, who died in 1881, and other family members and neighbors. A state historical marker, erected at the cemetery, recounts the "Limestone Cove Tragedy." Bells have continued to farm this land successfully through the decades, and James and David are names in the family's current generation.

The Bell Cemetery in Limestone Cove contains the graves of James Bell and eight civilians killed by Confederate troops in 1863.

Levi Trewhitt Farm
Bradley County

Character, reputation, and a lifetime of service to the community were often not enough to avoid harsh personal treatment during the war. Levi Trewhitt, Sr., purchased land in Bradley County in 1836 and served as one of the commissioners who laid out the county seat of Cleveland. He and his wife, Harriet Lavender, were the parents of fourteen children. Trewhitt and three of his sons, Levi, Jr., Andrew, and Daniel, remained staunchly loyal to the Union and advised their neighbors not to be a part of any rebellion.

Because of his vocal support of the Union, Judge Trewhitt was jailed in late 1861 during the Confederate response to East Tennessee's bridge-burner scare. Trewhitt and more than 200 others were arrested, but neither trial nor investigation of the charges was held before he was sent to a prison in Alabama. Family tradition holds that either Andrew or Levi, Jr. (sources name each one) was also imprisoned in 1863, first in Vicksburg and later Mobile, but escaped and returned to the farm where he hid for a time. In 1861, son Daniel had made his way to Kentucky where he was named a lieutenant colonel in the 2nd Regiment of the Tennessee Infantry Volunteers (USA).

After the imprisonment of Judge Trewhitt, a number of prominent Bradley County citizens, Confederate officials, and military officers signed a petition sent to President Jefferson Davis, noting the 65-year-old Trewhitt's advanced age, physical condition, and many years of community service. They asked for his immediate release, but before any action could be taken, Trewhitt died on January 31, 1862, still imprisoned, in Mobile. His remains were returned, and he was buried in Fort Hill Cemetery in Cleveland.

After the war, Daniel Trewhitt resumed practicing law and was appointed chancellor of the Chattanooga district, a position which he held until 1870. He afterwards served as a circuit judge. Andrew J. Trewhitt also returned to his pre-war profession and practiced law in Cleveland. He was elected district attorney general in 1868. In his later years, Levi Trewhitt, Jr., donated the land for the original Waterville Elementary School. He lived until 1916, and his great grandson owns the Trewhitt Farm today.

After the war, Andrew Trewhitt returned to his law practice in Cleveland.
Courtesy *Cleveland Bradley County Library*

Easterly Farm
Greene County

This East Tennessee farm was founded by John George and Mary Harpine Easterly in 1796 when they left their Virginia home and travelled to Greene County. Here they began to acquire land that eventually became one of the county's largest farms. Their 1,100 acres produced corn, wheat, oats, flax, and livestock. Son Jacob, married to Mary Bible, bought land of his own and also managed his parents' farm in their later years. This generation grew mulberry trees and cultivated silk worms, actually producing silk for a time, which their daughters wove into fine cloth. When his son, Abraham, died in 1842, Jacob gave 115 acres to his widow, Anna Parrott Easterly. With the help of her older son, Francis Marion, she operated the farm and raised her children.

Francis Marion Easterly acquired the farm from his mother and siblings in the early 1850s and continued to raise large crops of grain, cotton, and flax. Easterly became prosperous as a livestock trader by raising herds of horses, mules, and hogs, and then driving them over the mountains to the Carolinas to sell. In 1860, Easterly owned two slaves--Ann who was twenty-five and was purchased in 1856 for $700, and a fifteen-year-old male.

When the Civil War began, Easterly remained at home with his wife, Narcissa Powell, and their family of seven young children while two older sons joined the Confederate Army. Although Easterly was a Confederate sympathizer in predominantly pro-Union Greene County, the family apparently stayed on good terms with their neighbors and mutually aided each other during the hard times. The family, however, was brutally affected by the war.

One son, Frank, was killed while soldiering in North Carolina, and another son survived the battlefield but was murdered by bushwhackers on his way home. Similar roving bands of men, under no military control, were also considered by the family to be responsible for the death of Narcissa who died in childbirth in 1865 soon after she tried to keep "bushwhackers from stealing her woven materials hidden upstairs." Bushwhackers also kidnapped Francis and hung him from one of his own apple trees. After leaving him for dead, the outlaws moved on, but a female slave quickly cut him down and saved his life. The family maintains that the slave was Ann who, after she was freed, remained with the family for many years. She died at the age of 86 in 1920.

After the war ended, Easterly married Matilda Robeson in 1867, and they had six children. Family members attempted to put the past behind them and move forward with their lives, helping to rebuild the community. Realizing that education was the key to the future, Francis and four other local men founded the Parrottsville Academy in 1875. The opening of this institution finally "made it possible for even the poorest man to give his children an education." Easterly also gave money for a new building at Oven Creek Southern Methodist Church; the first building had been financed by his father, Jacob. Francis Easterly died in 1903. In recent years, descendants of Narcissa, Matilda and Francis Easterly, and those of Ann, who took the surname Easterly, have gathered at the farm.

F. M. Easterly survived hanging because his slave, Ann, cut him down after bushwhackers left him for dead.

Horner Farm
Perry County

A similar story of a near-hanging comes from Perry County, along the Tennessee River in Middle Tennessee. The area was divided in allegiance, and bands of hard-riding men from the two opposing sides crisscrossed the area while foraging, scouting, and skirmishing. The worst danger to civilians, however, was from the bushwhackers or "buggers" as they were known in the countryside of the lower Tennessee Valley region. The buggers were renegades from both armies, who joined with like-minded men to murder, torture, rob, and commit untold acts of mindless violence for their own purposes.

The outlaws and bandits respected no one's person or property. Amos Randel and his wife, Rebecca Finn, had come to Perry County from South Carolina to acquire fertile land and make a good living possible for their family. They acquired 490 acres along Cedar Creek in 1846 and grew a major corn crop to feed themselves as well as their hogs and cattle. The Randels also bought and sold tracts of land in the community over the years.

The 1860 federal census recorded that the 63-year-old Amos and 64-year-old Rebecca were living on their farm, assisted by two white laborers. During the war, a horrific incident occurred on the Randel place that could have affected the very future of the farm. According to the family, nightriders tried to rob the couple and tortured Rebecca Randel by repeatedly hanging her from a tree limb so that she would tell where their valuables were hidden. One of the hired hands had hidden in the bushes and saw that she was barely able to touch the ground with the tips of her toes. She dangled helplessly from the noose until the buggers finally left, and the man rushed to cut the rope. Fortunately, Rebecca's neck was not broken, but she never fully recovered from the trauma even though she lived to see the end of the war. She is buried nearby in the Flowers Branch Cemetery. In 1867, daughter Nancy Randel Horner and her husband, William Horner, inherited the farm. They had to sell 710 acres, but grew corn and peanuts, primarily, and raised livestock on the remaining acreage. Members of the fifth generation continue to grow crops and work the Horner Farm.

Gordon Farm
Rutherford County

The Gordon family, whose farm is within a few miles of the geographic center of the state, experienced some very difficult times during the war. On one occasion, the patriarch, John Hilton Gordon, was nearly hanged. Born in North Carolina, his marriage to Margaret White was recorded in Rutherford County, Tennessee, in August 1832. The Gordons had nine children and farmed near the west branch of the Stones River on over 100 acres by 1849. The Gordon family was opposed to slavery, but their sympathies were with the Confederacy and at least three sons fought for that army. Murfreesboro was occupied by Federal troops for most of the war, and evidence of an encampment is on the Gordon land. Because of this proximity and because of his allegiance, John Gordon and his family were under regular scrutiny. According to family history, Union soldiers once threatened to hang John in his front yard, possibly for information about the movements of opposing troops. Three times, they came to the point of carrying out their threat, but finally departed.

Amanda and A. W. Gordon survived the war, built a new home in 1870, and helped rebuild their farming community.

Amanda and her oldest grandson, Robert Jennings Gordon, Sr., are in front of the 1870 house about 1914, not long after the death of A. W. Gordon.

Alfred White (A. W.) Gordon, the second son of John and Margaret, was also a subject of Union attention. He and his wife, Amanda Nelson Gordon, of the Rucker community in Rutherford County, were married in 1860 and lived near his parents. On one occasion, A. W. Gordon was detained by Union troops for most of the day with the intention of persuading him to sign the oath of allegiance. As the day went on, members of the family became concerned about Alfred's ability to withstand the pressure. One of the Gordon relatives reportedly remarked, "They need not have worried; A. W. wasn't about to sign because he knew he would have to come home and tell his wife and Amanda Nelson Gordon would never have understood." Just after the war, in 1866, Alfred White Gordon purchased the farm that remains in the family. Like his father, A. W. was a farmer, but he also operated a cotton gin and wheat thresher for the benefit of the farming community. A. W. and Amanda built a home in 1870, and John and Margaret Gordon lived in a small house in the yard of this residence until their deaths in 1884 and 1888 respectively. The 1870 house was rebuilt in 1898-1900 to its present design. A. W. died in 1914, and Amanda lived until 1930. The current stewards, Gilbert and Ginny Gordon, maintain the farmhouse, the history, and the cemetery where seven generations of the family are buried.

LOYALTY OATHS

Because Tennessee was a contested state, control of cities, towns, and transportation routes sometimes alternated between Union and Confederate forces. A common practice of the controlling government was to require political prisoners and citizens to sign a loyalty oath or oath of allegiance. Sometimes the oath was a requirement for merchants who wished to remain in business, as was the case in Memphis after Federal soldiers occupied the town in 1862. As time passed, Memphians who wanted to make purchases also had to take the loyalty oath. Some Confederate supporters avoided taking the oath of allegiance to the Union, especially in the first years of the war. Historian Douglas W. Cupples recounts how some Memphis residents chose to leave their homes in 1862 rather than sign the oath of allegiance.

After the occupation of Nashville by Federal troops and the appointment of Andrew Johnson as military governor of Tennessee, the mayor and city council refused to take the oath of allegiance. Johnson retaliated by removing them from office. In Murfreesboro, army officials closed every church until the clergymen took the oath. During the summer of 1862 in Middle Tennessee, the Federal army rounded up prominent leaders, politicians, and ministers and imprisoned or exiled those who refused the oath. Many under Federal occupation in West Tennessee also faced the prospect of taking the oath. Gen. John Logan ordered all male citizens over the age of eighteen inside the picket lines in Jackson to take the oath of allegiance to the United States; those who failed to comply were arrested and imprisoned.

Confederate military leaders were just as adamant in requiring citizens to take the oath of allegiance to the Confederate government. By January 1862, Confederate officers were administering oaths of allegiance to officeholders in Knox and Sullivan counties. Elected officials in Cleveland, Tennessee, who refused to take the oath were arrested and imprisoned.

The oath of allegiance was not regarded by some men who took it as an iron-clad document. Some who had signed the Federal oath later might swear loyalty to the Confederacy when an area reverted back to their control, and vice-versa. Most of the fifteen people arrested near Humboldt, Tennessee, for destroying a bridge in July 1862 had previously taken the oath of allegiance to the United States; this oath did not stop them from planning and taking part in efforts to sabotage Union movements in that area. However, many of those who refused to take the oath in 1862 were taking it by 1864, when signs of Confederate defeat were everywhere, and citizens and soldiers, wearied by war, longed to stop fighting and return to civilian life.

The decision to sign the oath of allegiance was one undoubtedly faced by many farmers during the Civil War. Among them was William C. Rice, who acquired the Rice Farm in Wilson County from his father in 1864. He had signed the oath of allegiance to the United States on August 13, 1863. For many Tennesseans, whether Union or Confederate, loyalty oaths were part of the Civil War experience.

TAKING THE OATH OF ALLEGIANCE.

*Harris Mount purchased the 100-acre **Mount Farm** in Dyer County (now **Crockett County**) in 1860. He signed the oath of allegiance to the United States in 1863.*
From *J. T. Trowbridge, A Picture of the Desolated States; and the Work of Restoration*

Provost Marshal's Office,

6th Division, 16th Army Corps,

Fort Pillow, Tenn., Aug 26 1863.

I do Solemnly Swear, In the presence of Almighty God, that I will bear true allegiance to the United States of America, and will obey and maintain the Constitution and Laws of the same, and will defend and support the said United States of America against all enemies, foreign and domestic, and especially against the *Rebellious League* known as the Confederate States of America. *So help me God.*

H. N. Mount

CERTIFICATE:

Sworn and subscribed to before me, this 26 day of Aug 1863.

Ross Griffin
Capt. Co. G, 52d Ind. Inft. Vols. and Pror. Mar'l.

Residence _____

Age, 45 years; Height, 5 feet 1 inches;
Hair, light; Eyes, blue

Daniel's Dairy Farm
Dickson County

On old Highway 48, just north of the Dickson County seat of Charlotte, is an agricultural landscape of several farms that are characterized by nineteenth-century grazing fields, cemeteries, old transportation routes, and farm buildings. James Loggins and his wife, Nancy Grimes, established their homestead in this area in March 1861, three months before Tennessee seceded from the Union. Married in 1832, they were the parents of ten children. Not long after they moved to their new property, their son-in-law, Joseph Crawford Daniel, the husband of their daughter, Mary, rode up to the farm carrying his young son, William James. He brought the sad news that Mary had died at their home in Humphreys County. Daniel was preparing to join the Confederate army, so he brought little William to live with his grandparents. Thomas, the third child born to James and Nancy, also enlisted in the Confederate army in 1861, and he died at Ft. Donelson.

Back at the Barton's Creek farm, bushwhackers, in this case local renegades, came to the farm one day and dragged John Littleton Loggins, who was not yet old enough to join the army, from the house. While his family watched in horror, they tortured him by firing several shots at close range, making him think that the next bullet would kill him. The men finally tired of their game and left. Some years later, after John had grown into a very strong man, he was in Charlotte, recognized one of his tormenters, and confronted him. As his sister, Katherine Rebecca Loggins, remembered the incident, the two men fought in the courthouse yard until John was pulled off the man whom he was determined to kill.

Katherine Rebecca Loggins also recalled a time when she was returning on horseback from a friend's house and saw Union soldiers in the distance. She raced to alert her family and neighbors. Many of the women and children fled to a nearby cave where they hid while the soldiers continued on to their destination.

James Loggins died in 1884, having survived Nancy by six years. By the time of their deaths, their grandson, William James Daniel, was a young man and became the second owner of the acreage that carries his family name.

Generations of the family are buried in the farm's cemetery.

In the foreground is the remnant of a sunken road bed on the Daniel Farm that dates from before the Civil War.

Newman Farm

Jefferson County

The history of the Newman Farm attests to the virulent nature of civil war. John "Black Jack" Newman and his wife, Jane Caldwell Newman, acquired 100 acres in 1829 after moving across the mountains from Virginia. When the war came, Newman, a farmer and wagoner, and his family remained loyal to the Union. One son, Jonathan, joined the 9th Tennessee Cavalry (USA), and was later captured, and died in Andersonville Prison. In 1864, "Black Jack," who was sixty-eight years of age, was shot in the front yard of his home by Confederate soldiers and died from his wounds a few days later. The family recounts that Newman's surviving sons dared not visit their father in his final days or even attend the funeral because of fear of reprisal.

Camp Sumter in Andersonville, Georgia, like Camp Douglas, near Chicago, Illinois, was known for the inhumane treatment of prisoners. From Leslie's

Hendrix Farm
Cumberland County

Differing allegiances within a family could also lead to violence and tragedy, as happened in the Grassy Cove community. Settled by several families from Virginia, the cove is a lush and fertile area that includes over 10,000 acres with nearly half of that in bottomland. In a cove six miles long and two miles wide, descendants of the original families continue to own farms and practice traditional farming.

Dr. John Ford, Jr. and Nancy Loden Ford made their home on a farm founded by his parents in 1801. Dr. Ford practiced medicine but continued to operate the farm as well as the family store begun by his father. Four of John and Nancy's sons, Elijah, Thomas, John, and Elbert, served in the 1st Tennessee Cavalry (CSA) under Col. James E. Carter. Elijah (or "Lige") was captured and died in prison. A fifth son, Christopher A. Ford, who was a Methodist minister, married Elizabeth Swan, who came from a pro-Union family in Kentucky.

After Christopher left to join the Confederate army, his wife sued him for divorce and custody of their two young daughters on the grounds of desertion and nonsupport. He returned on furlough, counter-sued, and won the custody fight. He then returned to his company, leaving his children with his parents in Grassy Cove. On October 7, 1863, Elizabeth Swan Ford, along with her brother, Robert, and their father, came to the farm and murdered Dr. Ford as his wife, granddaughters, and his youngest daughter, Mary Jane, watched. The Swans then took the children to Kentucky. Mary Jane Ford was so affected by this incident that for a time she was unable to move or speak, and her mother was paralyzed "from fright."

When the sons of the Fords, based somewhere near Chattanooga, were notified of their father's death and the circumstances, they immediately rode to the cove along with some of their friends. On hearing the story, the accompanying soldiers, as well as some of the local residents, wanted to go after Elizabeth and kill the Swan family. The Reverend Ford begged them not to take this path of revenge, and he also feared that his daughters might perish in such a raid. After pleading with his comrades and praying before them, he was seized with a "violent lung hemorrhage" which left him bedridden for several days. He was unable to attend his father's funeral. He did recover, however, and apparently returned to his unit.

Grassy Cove was settled by several families in the early 1800s.

In the Grassy Cove Methodist Church Cemetery is the headstone for Mary Jane, John, and Rev. Christopher Ford near the graves of their mother and father. Just a few feet away is the burial site of Robert Swan, the alleged murderer of their father.

An account by John Ford, Christopher's brother, describes his own return to the farm after the war. He found Nancy still paralyzed and his sister, Mary Jane, very frail and the sole caretaker of the farm and their mother. Though John had planned to marry after the war, he decided that he could not bring a bride to a place that was so devastated and lacking in any comforts, and where she would also have the burden of caring for his mother and sister. Ford assumed the responsibility for his mother and his sister until their deaths in 1874 and 1900 respectively.

Several of the people who played a part in this drama returned to Cumberland County after the war. The 1870 census shows that Elizabeth Swan was living with her parents, Margaret and Thomas Swan, in nearby Crossville, along with her daughters, Margaret (age 12) and Lilly (age 10). Local history recounts that, soon after the war, Elizabeth sought out her former husband in hopes of reconciling with him, but he refused her. Elizabeth remarried as she is later listed with the last name Glover. Elizabeth's brother, Robert Swan, who is presumed to be the one who actually killed Dr. Ford, remained in Pulaski County, Kentucky, at least through 1870, but returned to Grassy Cove, where he is listed as a farmer in the 1880 census. Rev. Christopher A. Ford resumed his pastoral duties and served at the church his grandfather had founded in the cove. He lived on the farm with his siblings, Mary Jane and John, and shares a marker with them in the Grassy Cove Methodist Church Cemetery. In the same cemetery are tombstones for Dr. John Ford, Nancy Loden Ford, and Robert Swan.

Dr. John Ford, left, photographed sometime before October 1863, was murdered by his daughter-in-law's family. Christopher Ford was married to the woman whose family was responsible for the murder of his father.

PHOTOGRAPHY IN THE CIVIL WAR

Photography was still a relatively new art form in 1861; however, photography and photographers soon had an enormous role to play in the war and on the home front. Soon after President Abraham Lincoln called for 75,000 troops to put down the rebellion, photographer Matthew Brady acquired permission to follow and photograph the troops. Brady's foray into the thick of battle was short-lived because he lost his equipment in the tumult. Soon, however, he began to send photographers into the field to capture the grim aftermath of the battles. The techniques (wet-glass negatives or collodion-on-glass) used by photographers at the time were laborious and could not produce action shots. Instead, photographers, including Alexander Gardner, Timothy O'Sullivan, and others shot camp scenes, strategy sessions, preparations for retreat and the results of battles, including scenes showing the destruction and unbelievable number of casualties. The Civil War was the first time that the American public saw actual images of dead soldiers on the battlefield, rather than sketches drawn by artists.

In addition to battlefield photographs, portraits were also extremely popular with soldiers and their families at home. Tintypes were less expensive to produce than daguerreotypes, and itinerant photographers roamed the camps, posing soldiers against painted backdrops. These images were then sent to loved ones back at home. Soldiers often carried with them tintypes and daguerreotypes of family members who had posed specifically for that purpose..

Generations of family members on **Oakdale Farm** in **Cheatham County** have studied this image of their relative, William H. Smith, who was nicknamed "Button." Button's great-grandfather and farm founder, Charles Gent, fought for the colonies in the Revolutionary War. Button Smith was among the Confederate soldiers captured at Ft. Donelson in 1862, but he was exchanged later that year. He died in the Battle of New Hope Church, Georgia, in 1864. This tintype of Smith, in uniform with rifle and knife, was a popular pose for Union and Confederate soldiers.

Several photographs of Capt. David Preston Sherfy are kept by his descendants on their farm in **Washington County**. He first joined Company H, First Illinois Cavalry, and saw action at Ft. Henry, Ft. Donelson, and Shiloh. In March 1864, he became an officer with the 3rd U. S. Colored Cavalry. More about his service and family are recounted in the entry on the **Michael Krouse Farm** in "The Struggle to Farm."

Identified as Cavalryman Jackson of the 3rd United States Colored Cavalry, this trooper was detailed to assist David Sherfy after his horse fell on him and broke his leg in February 1865. The 3rd U. S. Colored Cavalry, made up of white officers and black troops, was ranked as one of the finest cavalry regiments in the Army of the Tennessee. Jackson's photograph is also part of the collection of David Sherfy's descendants.

Lone Pine Farm
Washington County

Even after military actions ceased, peace was slow in coming. Violence continued in areas like Knob Creek, just outside present-day Johnson City, where neighbors and families were zealous in their allegiances. James Crumley and his wife, Elizabeth Caroline King, experienced the effects of conflict firsthand in their home. Crumley, who purchased his farm on Knob Creek from the Krouse family in 1852, was an enrolling officer for the Confederate army. His wife's brother, Alfred J. King, was also a known Confederate sympathizer. In September 1865, William B. King (the relationship, if any, to Alfred J. King is unknown) of Carter County asked fellow Unionist Lafayette Miller, also of Knob Creek, to help him and his men capture Crumley and King. Apparently, as the family recalls, a sizeable reward of $5000 was being offered for the two men, though it is unclear who was offering this sum following the war.

Miller, age 23 and a former quartermaster in the 8th Tennessee Cavalry (USA), had been captured at Russellville on November 13, 1864, and confined in Morristown before returning to his unit in March of the following year. Miller's family had been divided over the war; his brother, George W., and his son had joined the Confederacy, while other family members remained loyal to the Union. Another brother, Alfred Miller, had been killed by Confederates while returning to his home after feeding his livestock. Mustered out of the Union army in Knoxville, Lafayette had returned to his home just a few days before hearing from William B. King.

Lafayette Miller agreed to accompany King and others to the home of the Crumleys on September 28. When they arrive, both Crumley and Alfred J. King were hiding in the attic of the house. Crumley ran from the place, leading William King and his men on a chase. Miller and A.J. King fought in the attic, down the stairs, through the house, and out into the yard. Apparently, King fatally wounded Miller, but not before being seriously wounded himself. When William King returned after capturing Crumley, the group is said to have "finished off" A.J. King, who lay in the yard. A.J. King was interred in the White Family Cemetery in Johnson City, and tradition has it that Lafayette was buried in the Miller Family Cemetery on Knob Creek, though his grave is unmarked. James Crumley lived until 1883 and is buried in the family cemetery. Crumley descendants still own the farm and the house where this incident took place.

Generations of the Crumley family, and their horses, patiently waited for their image to be taken in the 1880s. The house was the scene of violence just after the war.

W. F. Collins Farm
Carroll County

This is the only known image of Daniel Collins, blacksmith and gunsmith, who was killed by bushwhackers in 1864.

These powder horns and the other items are part of the family's collection of tools used by the skilled smithy.

The violence that characterized the Civil War also shaped the history of the Collins family farm in West Tennessee. Daniel Collins and his wife, Miranda Lett, moved from Virginia to establish a farm southwest of Clarksburg on the Rutherford Fork of the Obion River in 1849. In addition to farming, Collins worked as a blacksmith and a master gunsmith. From the 1830s until his death, Collins kept a small bound book in which he recorded precise instructions for his blacksmithing work and also such details as how to engrave letters on guns. He listed "blacksmith" as his occupation in 1850 and "gunsmith" in the 1860 census. Either or both of these allied professions made him a useful man, or perhaps a targeted one.

Early in the war, Carroll County was controlled by the Confederate army, though many citizens remained loyal to the United States. Guerrillas and bushwhackers were a terror to the people of the county, who suffered much more from the depredations of these renegades than from the actions of either army. Roving bands of outlaws committed several brutal murders within the county.

The Collins family, whose sympathies are unknown and who may have tried to remain neutral during the war, endured a fateful incident in which a band of armed men rounded up a group of local citizens, including Daniel Collins, and confined them in a schoolhouse. He escaped, hid in the woods, and eventually returned home. On June 8, 1864, some of the same men returned, found Collins, and killed him. A brief will made and signed by Daniel Collins is dated the day of his death, indicating that he survived long enough to deal with the disposition of his property. In later years, his descendants discovered this poignant epitaph, obviously written by someone who knew Collins well, inserted at the back of his worn manual of metalwork:

Wonct this book a friend did own
Who could play a hamer or a twirl a drill
With skill ingenuity and grace
Surpassing all the knowing smiths afar
Small and spunky tru and firm
He'd lived to see that bloody war of 61
Alas how many have fallen
O that he had lived to see the peace of 66.

END NOTES

Officer Farm

West, Carroll Van, Mark Grindstaff, and Sean Reines. "Officer Farmstead and Cemetery, Overton County, Tennessee." National Register of Historic Places Nomination Form. Tennessee Historical Commission. Listed 2001.

Bell Farm

Scott, Samuel W., and Samuel P. Angel. *History of the Thirteenth Regiment, Tennessee Volunteer Cavalry, U.S.A.* (Philadelphia, P. W. Ziegler & Co.,1903), 339-340.

Bell, Sandra, contributor. *Bell Cemetery Records, Limestone Cove, Unicoi, TN* (USGenWeb Archives, 2007).

Gordon Farm

Gordon, Thomas Gilbert, Sr., MD. *The John Hilton Gordon Family.* Rutherford County, Tennessee, (n.p,1988).

Daniel's Dairy Farm

Loggins, Joe. "Loggins in Dickson County - The Family of John Littleton Loggins," *The Heritage of Dickson County, Tennessee, 1803-2006* (Dickson County Heritage Book Committee and County Heritage, Inc., 2007), 297.

Trewhitt Farm

Allen, Penelope Johnson. "Leaves from the Family Tree ….Trewhitt," *The Chattanooga Times, Chattanooga,* June 24, 1934.

Horner Farm

Bowen, Mary (Sewart). "Randel Cemetery," *Perry County, Tennessee Cemetery Records* (n.d).

Hendrix Farm

Harvey, Mrs. Stella Mowbray. *Tales of the Civil War Era* (Crossville, Cumberland County Civil War Centennial Committee, 1963), 22-28.

Stratton, Cora S. And *This is Grassy Cove* (Crossville: Chronicle, 1938).

http://fordgen.tripod.com/civilwar.htm

http://genforum.geneaology.com/ford/messages/282.html

Lone Pine Farm

Sell, Jeanne Lyle. *The Miller Family of Knob Creek, Washington County, Tennessee* (n.p. 1985), 174-175.

Insets

Loyalty Oaths

Ash, Stephen V. *Middle Tennessee Society Transformed 1860-1870: War and Peace in the Upper South.* (Baton Rouge: Louisiana State University Press, 1988).

Cupples, Douglas, W. "Rebel to the Core: Memphis' Confederate Civil War Refugees," *West Tennessee Historical Society Papers 51* (1997), 74-73.

Fisher, Noel, C. *War at Every Door: Partisan Politics and Guerilla Violence in East Tennessee, 1860-1869* (Chapel Hill: University of North Carolina Press, 1997.

Foner, Eric. *Reconstruction: America's Unfinished Revolution 1863-1877* (New York: Perennial Classics, 1988).

Murran, Nathan K. "Military Government and Divided Loyalties, The Union Occupation of Northwest Tennessee, June 1862-August 1862." *West Tennessee Historical Society Papers 48* (1994): 91-106.

Photography

Newhall, Beaumont. T*he Daguerreotype in America*. (New York: Dover Publications, 1975).

Holley, George W. *David Preston Sherfy, Kinsman and Cavalryman.* (Knoxville:TN, Earthtide Publications, 1986).

"Photography and the Civil War, 1861–1865." *Heilbrunn Timeline of Art History.* http://www.metmuseum.org/toah/hd/phcw/hd_phcw.htm.

Zeller, Bob. *The Blue and Gray in Black and White: A History of Civil War Photography* (Westport, CT: Praeger Publishers, 2005).

Refer to the "Bibliographical Essay" for additional resource materials.

The War and Work Animals

Animal power, supplied by horses, mules, and, in lesser numbers, oxen, was an essential part of any farming operation before the Civil War. With just one animal in reasonably good physical condition, a farm family could clear land, make a crop, haul surplus to market, and travel. Otherwise, farm work was limited and life in general was much more difficult. The U. S. Agricultural Census of 1860 reported a total of 6,115,458 horses, of which 1,698,328 were recorded in states that would secede. Mules numbered 800,663 in southern states with only 328,890 counted in other states. In that same year, the price of a horse was about $125 and a mule was somewhat less. Because horses and mules were a necessity to armies, they were constantly in demand, and their confiscation from farms during the war resulted in serious hardships for families.

Horses were the backbone of military operations. They were indispensable to the cavalry, of course, but equines, mules, and oxen were also the main source of power for "wagon trains" that hauled food, supplies, cannon, guns, and ammunition. Therefore, both armies required a steady supply of animals. Hundreds of thousands of these four-legged soldiers served each army, and it is estimated that more than one million died, while countless others were wounded.

As the war continued, the need for replacement horses and mules became more challenging. By 1863, the Union army reportedly needed 500

(Opposite page) Horses and mules were essential for moving supplies and artillery.
From *Harper's*

An estimated one million horses and mules were killed and thousands more injured during the Civil War.
From *Leslie's*

new horses each day to replenish its stock. Especially in the northern states, breeders who supplied horses and mules prospered during the war. For Confederate forces, who were expected to supply their own mounts, securing horses and mules to ride and serve as pack animals became increasingly difficult.

The most widely used breeds in armies were the American Saddlebred, Tennessee Walker, and Morgan, but the military also took carriage and riding horses of no particular breed and even ponies. The armies "enlisted" mules, known for their strength, sure-footedness, and endurance. These hybrids, a cross between a mare and a male jackass, were used mainly to carry supplies and haul cannon. Generally prone to panic and stampede under fire, mules were harder to control and had to be kept away from battle as much as possible. Both armies constantly foraged for feed and scouted for water sources for their animals.

After the war, replenishing the hundreds of thousands of animals lost during the fighting was a boon to the state's agricultural economy. By 1870, the average price for a mule was about $106, while a horse could be expected to bring about $85. Tennessee's reliance on animals continued until the mid-twentieth century when tractors and other machinery became more affordable and popular with farmers. Horses and mules are still a common sight across the Tennessee landscape, and farmers continue to breed, show, sell, and sometimes work the land with these animals whose history is so much a part of the Civil War and the rebuilding of the state.

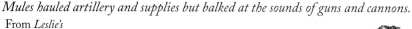

Mules hauled artillery and supplies but balked at the sounds of guns and cannons. From *Leslie's*

THE MULE CORRAL AT PITTSBURG LANDING.

(Opposite page) Mules were so much a part of life in Tennessee that musical compositions about the hybrid animal were popular. This piece, published in Nashville in 1862, was owned, and presumably played, by Nora Gardner of Weakley County.
Courtesy *Center for Popular Music, Middle Tennessee State University*

HERE'S YOUR

(Found at Last!)

MULE GALLOP!

Composed by

CHARLES STEIN.

PUBLISHED BY **C. D. BENSON**, NASHVILLE, TENN.

NASHVILLE:	MEMPHIS:	NEW ORLEANS:	MOBILE:
J. A. McCLURE.	J. A. McCLURE.	A. E. BLACKMAR & BRO. B. P. WERLINE,	BROMBERG & SON. J. H. SNOW.
GRIFFIN, Ga.: BROWNER & PUTNAM.	MACON, Ga.: J. W. BURK.		EUFAULA, Ala.: JAMES. A. HUNTER.

Allison Farm
Washington County

The Allison Farm in Washington County, founded in 1800 by John Allison II, experienced nearly constant troop movements as armies from both sides passed to and fro along the East Tennessee & Virginia Railroad that ran through nearby Jonesborough. Within five miles of the farm is a county road, known as "Yankee Camp Road," because of the detachment of Union soldiers stationed there for most of the war. E. S. Morrell, the third owner of the farm, enlisted in the 1st Tennessee Cavalry, Co. A (CSA). Before leaving the farm, however, he and his wife, Susan Allison Morrell, granddaughter of the founder, devised a plan to keep their four draft horses safe. At the farm's blacksmith shop, a huge chain was forged with "links as big as a man's hand," along with a huge padlock. The purpose of the chain and lock was to keep the horses secure in a log barn at night. The chain passed through holes in chinks on either side of the door. Once the horses had done their work, they were taken directly to the barn, where the owners took corn, hay, and water to them daily. In this way, the Morrells successfully kept their horses safe during the years when so many families had their livestock confiscated by troops or stolen by brigands. The family continued to keep draft horses until the 1960s, and the story of the "Horse Barn Stable" has been passed down by each generation.

The log portion of the barn which sheltered the Allisons' horses during the Civil War, between the first and third poles from the left, survived until 1978 when it was burned by vandals.

Cartwright-Russell Farm
Smith County

Richardson Cloud Cartwright, a descendant of early settlers who came with John Donelson to establish Fort Nashborough (Nashville) in 1780, purchased 226 acres from his father in Smith County in 1832. Cartwright married Henrietta Dean in 1840, and they were the parents of seven children, ranging in age from one to thirteen years, when he died of pneumonia in 1854. While Henrietta was in better financial shape than many women, transacting business was difficult for females in the 1850s. She managed the farm and raised her younger children with the help of her eldest son, Clark, who was born in 1842. Henrietta Cartwright was especially known for breeding fine-blooded horses.

When the war erupted, so did dissension within the family. Clark joined the Confederate army and served as a corporal in Co. H, 28th Tennessee Infantry, while his brother Hailey, only fifteen, enlisted at Carthage and fought for the Union in Company D, 8th Tennessee Mounted Infantry Regiment. While the two older sons were away, Henrietta and her five younger children were wary of the guerillas who often roamed Smith County and the Cartwright community. The family recorded how Henrietta took special care to hide her horses in an old log barn some distance from the house. On one occasion, though, the soldiers found and took the horses. Through some inquiries, Henrietta learned her horses had been seen in Kentucky. She left immediately to try to reclaim them and, against all odds, succeeded in locating her horses and bringing them back to the farm.

Brothers Clark and Hailey Cartwright survived the war but never overcame their differences, which often surfaced over the years. Hailey and a third brother, John, became physicians and moved away, while Clark remained on the farm. Amanda Cartwright, the middle child, was a successful seamstress, often travelling as far away as Texas to sew for other families. Her three sisters married. The descendants of the youngest of the seven children, Elizabeth Cartwright, who married Leonidas (Lon) Shelby Russell, own the farm today. They continue to pass along the stories of Henrietta, their matriarch and a determined and successful horsewoman.

Henrietta Dean Cartwright (seated), and two of her children, Amanda and Clark, with one of their horses, worked the farm after the war.

Burrow-Gregory Farm
Macon County

William S. Burrow and his wife, Elizabeth Ford Wilson, and their 10 children lived on a farm that had been established at least by the 1830s. The Burrows owned slaves and some of their history remains with the family. Documents include a bill of sale from 1852 that records William Burrow's purchase at auction of "a negro boy named George about Eight years old, for the sum of five hundred and fifty three dollars." In 1858, William Burrow bought Roda and her child, Hannah, for $1126.50. In the 1880 census Roda and two children remained as domestic servants in the Burrow household.

The Burrows raised and worked both horses and mules to cultivate tobacco and corn crops. The family tells the story of a time during the Civil War when William and his sons were plowing their mules and Union soldiers took the animals out of their harnesses, leaving the men standing in the field with the plows. Another story, perhaps from the same time, involves a soldier asking William Burrow if a particular horse could be ridden. Burrow assured the man that it could be, but when the soldier mounted the animal, he was immediately thrown. This result "came very close to getting the owner shot on the spot," wrote descendant Jack Burrow Gregory.

Securing fresh horses and mules, along with food for armies and livestock, depleted farms across the state.
From *The Soldier in Our Civil War*

Hunter Cove Farm
Putnam County

Putnam County's Dudley Hunter was a well-known horseman. His wartime effort to save a favorite mount ended tragically with consequences that affected his immediate family for years. Hunter's mother was Sarah Boone Hunter, a niece of famed frontiersman Daniel Boone, who had travelled with her uncle across the mountains. She married William Hunter in 1787, and they purchased Hunter Cove Farm in 1814. The family and their slaves raised timber, corn, oxen, and swine, as well as horses. Son Dudley acquired about 5,000 acres in 1848. After the death of his first wife, Polly Clark, Hunter married Amy Gist Lowery. The Hunter family lived in a one-story brick home constructed by their slaves, which included thirteen men, women, and children in 1860. Hunter was particularly fond of his horses, which he bred and trained, and he also built a race track on his farm.

Hunter provided numerous horses to Confederate cavalrymen when the war began. In October 1862, he was on the square in Sparta when Texas Rangers came to town looking for fresh mounts. When the soldiers tried to seize his animal, he told them he would give them any horse or as many horses as they wished if they would leave his favorite colt alone. The Texas cavalrymen killed Hunter because he refused to give up the animal.

Amy Hunter, left with the children and a large plantation, did what was necessary during the war years. Some of the former slaves remained after emancipation and with their help and that of her mother-in-law, "Granny Boone" (who lived until 1866), she was able to maintain the property. A family story recounts that during the war, Amy drove an oxcart to Louisville on the Old Kentucky Road to exchange and barter what she could for supplies and food. Her sons, Rush, born 1854, and Vance, born 1859, were too young to serve as soldiers. As the years passed, they lived and worked on the farm, married and built their own houses, and acquired the property in 1883 after their mother's death. The Hunter brothers became known for the herds of Missouri mules they raised and trained for resale. The history of the Hunters' farming operation is interwoven with horses and mules.

Union and Confederate cavalry units required thousands of horses during the war.

From *Leslie's*

Isaac Huddleston Farm
Putnam County

Most families were unable to keep their animals safe for any period of time, and many owners never received compensation for livestock taken by soldiers. One exception was James S. Robinson who was noted for raising some of the finest mounts on the Upper Cumberland Plateau. By 1860, Robinson and his wife, Syrena Isom, were operating the farm founded by his parents in 1841. Family tradition holds that Union troops confiscated James Robinson's horses but they also took him prisoner to care for the animals. In 1881, the United States government awarded Robinson 890 acres, with an additional 57 acres in adjacent White County granted in 1887, as compensation for his horses. James and Syrena Robinson lived for a number of years on their Hickory Valley property in White County. Their daughter, Mary Jane and husband, Asbury Bullock, continued farming the original Putnam County farm after Syrena's death in 1914. Several generations later, the working farm remains in the family.

James S. Robinson was one of the finest horsemen in the Upper Cumberland region. He posed on one of his mounts in the 1880s.

Syrena Isom Robinson, photographed about 1900, lived until 1914.

Davis Farm
Henderson County

Asa Davis and his wife, Annie Wood Wilkinson Davis, were the parents of nine children. Asa, a veteran of the War of 1812, sold 420 acres to his son Columbus Marion Davis in 1855. The younger Davis and his first wife, Martha Jane Breazeale, also had nine children. In *Henderson County,* G. Tillman Stewart writes that Columbus Davis of Browns Creek was noted for "fine livestock, excellent corn and wheat, and soil conservation practices." Describing the conditions of farming in Henderson County following the Civil War, Stewart wrote, "Both armies had confiscated virtually all good horses and mules in the county; consequently, it took nearly five years to replace stock to enable agriculture to begin again. Good breeding animals frequently were imported from the North to aid this effort: Columbus Davis owned expensive mules; Davis' jack was imported from Spain." Also active in the political rebuilding of the county after the war years, Columbus Davis, whose second wife was Arista Hare, was justice of the peace for thirty nine years.

Columbus and Arista Hare Davis relied on mule breeding and sales to help regain financial stability after the war.

Austin Farm
Lauderdale County

Travelling by covered wagon from North Carolina across the mountains at the age of eight was the beginning of Louisa (Louiza) Eliza Castellow Austin's long life in West Tennessee. Her parents were Henry D. and Sarah Spruiell Castellow, and, in April 1855, after arriving in Lauderdale County, her father purchased 160 acres. He died in December of that year, leaving Sarah to raise their five children, manage the farm, and pay the debts and taxes on their new property. To make these payments, Sarah Castellow sold 40 acres in 1856. Sarah and her children, Benjamin, Mary Ann, Jane E., Louisa (Louiza), and Henry, raised vegetables, had a milk cow, chickens and hogs, and worked their cotton crop with horses. Louisa, the eldest child, married John Richard (J. R.) Austin in August 1863 when she was 21. Austin, age 26, had served in Co. G., 4th Infantry (CSA), but was discharged in January 1863 "by reason of disability." Benjamin Castellow was also in the Confederate army. Sarah and Louisa, along with Henry, who was too young to be a soldier, and J. R. Austin, suffering from consumption, managed the farm.

During the latter months of the war, Union soldiers burned the Castellow's cotton crop and took all of the horses except one. Sarah Castellow successfully pleaded with the soldiers to leave one animal so that she and her family might be able to make a new crop. The family had a very difficult time during and after the war. Benjamin Castellow died in 1870, and the farm was sold at public auction in November of that year. J. R. Austin, whose health continued to decline, died in 1873. Louisa Castellow Austin took charge of the family's circumstances, and in November 1880, almost to the day when it had been sold a decade earlier, she bought 120 acres of her former family farm from P. T. and Susan Glass. This was the original acreage bought in 1855 by her father, Henry Castellow, less the 40 acres sold by her mother, Sarah. Louisa Castellow Austin raised her five children and deeded the farm to her eldest son, Charles Thomas Jefferson Austin, in 1911. She remained on the farm until her death in 1929. Her great-granddaughter, Susan Austin Wieber, and her family manage and work the same acreage that Louisa returned to her family.

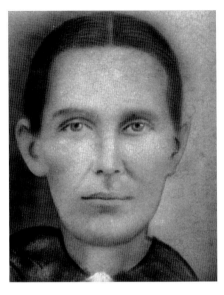

Louisa Castellow Austin, photographed about 1870, cared for her disabled husband, their children, and the farm.

Louisa Castellow, center, is surrounded by her family in the 1890s.

Keller Home Place
Lauderdale County

Few veterans were able to find a horse to help them return home after the fighting ceased, but Hiram Washington Keller managed to do just that. Keller and his wife, Roberta Burks Keller, purchased a farm of 118 acres just east of Henning in Lauderdale County in 1861. A sergeant with the 7th Tennessee Cavalry (CSA), Keller was captured during the siege of Vicksburg in 1863. After being released from prison, he looked for the thinnest horse he could find to ride home. He wanted one that looked so poorly that no one would want to steal it. He and the horse made it back to his farm.

About 1868, the veteran built a house for his wife and several children which, though remodeled and updated through the years by the family, remains the primary residence today. Active in the rebuilding of their community, the Kellers were founders of Bethlehem Methodist Church and the local Henning Academy. H.W. and Roberta managed a diverse farm. Roberta died in 1899, and Hiram lived until 1928. The couple provided a good foundation for a successful and progressive operation that continues today with their descendants on the Keller Home Place.

H. W. and Roberta Burks Keller married in 1860 and established their farm the following year.

Foulks (Hutchinson) Farm
Obion County

The building of railroads before the war, and their rebuilding after the war, provided an economic boon to those who could supply mules for the work. In the late 1850s, John J. Foulks of Obion County helped to build the New Orleans & Ohio Railroad. He supplied logs from his farm for cross ties and hired out teams of mules. His great-granddaughter, Martha Allen Hutchinson, tells the story of a huge iron chain the family owns. In the winter of 1857, "the weather was so bad the men could not get the logs out of the bottoms and therefore fell behind on orders." Railroad officials wrote to John J. Foulks to inquire about the delay asking, "Are your mules too weak?" Foulks replied, "I have 40 teams of mules that could pull your trains if the chain was strong enough." The next week, Foulks received a massive chain, work proceeded, and tracks were back on schedule for the line's completion in 1858.

In May 1866, Foulks and his wife, Elizabeth, acquired the deed to 350 acres just south of Harris Station on the rail line. Foulks continued to cut timber from his land for cross ties and breed, sell, and work mules as the New Orleans & Ohio was consolidated with other lines to support the expanding agricultural, industrial, and commercial markets of the latter part of the nineteenth century.

This massive chain, forged in in 1857 by the New Orleans & Ohio Railroad, was used by John J. Foulks and his mules to pull logs out of muddy bottom land to build cross ties for the tracks.

*George Washington Mansell of **Putnam County** operated a blacksmith shop for the community. He and his family raised a variety of grains for their own use and to feed their mules. The tradition of raising and working mules continues with the present-day owners of the **Mansell Farm**.*

END NOTES

Introduction

Grace, Deborah. "The Horse in the Civil War," *Horses of the Civil War.* http://www.reillysbattery.org/Newsletter/Jul00/deborah_grace.htm

Stewart, G. Tillman. *Henderson County* (Memphis State University Press, 1979). http://www.tnyesterday.com/books/stewart/1866-1890.html

Refer to the "Bibliographical Essay" for additional resource materials.

Cleburne Farm, Maury County, p. 132

Matt Gardner Homestead, Giles County, p. 117

McQueen Farm, Loudon County, p. 38

Fermanagh-Ross Farm, Greene County, p. 5

Cypress Creek Farm, Benton County, p. 150

Woodard Hall, Robertson County, p. 139

Samuel Raulston Farm, Marion County, p. 132

Grassy Cove, Cumberland County, p. 57, 101, 102, 105

Elmwood Farm, Rutherford County, p. 27

Tilson Farm, Unicoi County, p. 107

Fairview Farm, Jefferson County, p. 40

Dover Hotel, Surrender House, Vaughn Farm, Stewart County, p. 18

The Struggle To Farm

Farming during the war was nearly impossible in many parts of Tennessee. Troops were constantly on the move, and no farm was safe from foraging. Men of soldiering age left to join the armies, and women necessarily assumed even more of the daily and seasonal work. Many slaves left for places of freedom as soon as they could, some to take up residence in contraband camps which offered protection and food behind federal lines. Other former slaves joined the Union Army. Their absence immediately changed the scope of work on farms that had previously relied on the labor of these men, women, and children. Along with the responsibility of caring for family members young and old, women managed farm work either on their own or with the help of their children, relatives, neighbors, and sometimes former slaves who chose to stay after emancipation. As the fighting continued, soldiers took leave to head home to help plant, harvest, and hunt game whenever possible. They could, however, be absent for months or years. Ultimately, death on the battlefield or in prison, disease, or infected wounds claimed thousands of lives. In many families, returning veterans were so disabled from injuries that they were unable to resume rigorous farm labor.

Farm families throughout Tennessee regularly received unannounced visits from the fighting legions foraging for food, grains, animals, guns, tools, clothing, boots, or anything useful. Union and Confederate forces, as well as guerillas and bushwhackers, constantly scoured the countryside for supplies to keep men and work animals fed and on the move. Soldiers dismantled fences and removed or burned them in place for heating and cooking campfires. In some cases, troops took apart whole farmhouses and outbuildings, carted them off for other uses, or set them on fire.

Farms located near major roadways, waterways, and railroads were more frequent victims of raids, though homesteads in more isolated areas of the state did not escape the loss of their moveable property. Time and again, farm families had little recourse in the face of armed men who took what they wanted and, more often than not, left without providing payment or even a receipt. As each year of interminable fighting continued, the situation deteriorated; too many farm families were left without adequate provisions to feed themselves and the means to plant a new crop or replenish their livestock.

In spite of the hardships and dangers, those at home did their best to continue the traditional cycles of planting and harvesting crops, breeding and raising animals, and preparing and storing food. With a determined will, coupled with a good measure of stubbornness, craftiness, or sheer bravado, families from the Mississippi River to the mountains survived and kept the farms going.

To sustain armies, foraging was a common practice. Farms were targeted time and again until nothing was left for the farm families.
From *Leslie's*

Corona Farm
Tipton County

Federal forces controlled the Mississippi River, posing a direct threat to those who owned valuable land along the waterway, the vast majority of whom supported the Confederacy. Devil's Elbow, the sharpest bend between the Ohio River and New Orleans is located near Island #37 in Tipton County. The war directly affected this beautiful place where naturalist John J. Audubon once visited.

At the time of the Civil War, Devil's Elbow was also the name of the plantation that John and Elizabeth Bradley Trigg had established in 1836. The family's main residence, however, was just south of Memphis at Fort Pickering, a frontier-era fort re-established by the Confederates in 1861 and expanded by the Union after occupation of the area began in June 1862. Trigg, a city alderman, held interest in three railroad companies, was a director of the Planter's Bank, and owned rental property in Memphis. At his death in 1863, his property included nearly 16,000 acres in three states.

Trigg's estate settlement was complicated by the war. At her marriage in 1846, Lucy, the daughter of John and Elizabeth, was given a sizeable portion of Island #37 and renamed it "Corona." Her husband, Charles Ambrose Stockley, was in the Confederate army, and Lucy and her children were evicted from the Fort Pickering house by the Union forces occupying Memphis. Devil's Elbow, twenty-five miles away, took several hours to reach no matter what the mode of transportation. Also, it was difficult to haul sufficient supplies to the island because a Federal pass was required to transport even limited quantities of foodstuffs and other items.

Nevertheless, Stockley loaded her family and what she could take into a wagon, made the journey to Corona, and managed the farm for the duration of the war. Finding and keeping a labor force on the island was challenging. Further, two of Lucy's sons were murdered – one by bushwhackers and the other at the commissary on the plantation. Family history recounts that her Fort Pickering residence was burned after President Lincoln's assassination when Lucy, whose sympathies had always been with the Confederacy, objected to the house being draped in black.

The Mississippi River made its own change to the farm by violently and suddenly altering its course in 1876. In just two days, more than 1,000 acres of Corona were lost. Since that time, the land, though officially part of Tipton County, has actually been on the Arkansas, or west, side of the river. The ever-changing landscape, dominated by the Mississippi River, and the unique heritage of the Trigg and Stockley families, is remembered by their descendants who continue to live on Corona.

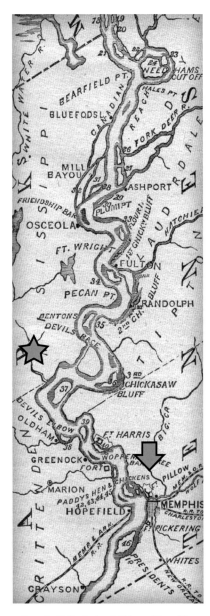

On the Mississippi River, Devil's Elbow, Island #37 (see star), is about twenty-five miles from Ft. Pickering, which is just south of Memphis (see arrow).
From *Harper's*

Light House Farm, Smith Farm, and Gauldin Farm
Dyer County

Dyer County, with the Mississippi River as its western boundary, saw no major battles during the war, but both armies regularly operated in the area as did guerrillas. The rich farmland was subject to constant foraging, and the families were hard pressed to continue raising row crops and livestock.

Joel A. and Sue Light established their 225-acre farm just west of Dyersburg in 1852. They purchased their land from Benjamin McCullough, a future Confederate general who had served under Gen. Zachary Taylor in the Mexican War. The Lights grew tobacco, cultivated orchards, and raised hogs and horses with the help of slaves. In the census of 1860, they owned twenty-seven slaves. The Civil War and its aftermath were "shattering times," recalls the family. Troops visited the farm, and several slaves left to join the Union army and live at contraband camps. The family's determination to keep their possessions from being confiscated led them to bury their valuables and even go so far as to drive a nail into the hoof of a prize stallion so he would not be taken. Joel Light died in 1880 and his widow, along with their five children, inherited the farm which they soon changed from row-cropping to raising cattle. The stories of the family, along with a pre-Civil War house, are preserved by the current generation at Light House Farm.

West of the Friendship community in Dyer County, Simon Peter Hawkins and his wife, Isabella Taylor, established their farm in 1852. Their 100 acres, bought for $3 per acre, supported a diverse agricultural operation that included timber, row crops, vegetables, and fruit trees, along with dairy and beef cattle. Hawkins enlisted in Capt. William Gay's company of the 47th Tennessee Infantry (CSA) at Trenton in March 1862. After being severely wounded in combat, he died in an Atlanta hospital in July 1863.

Isabella became the owner of the farm and was responsible for its management and for raising ten children. Her success as a mother and farmer is remembered by her descendants and the community. After her death in 1892, J.W. Smith, who had married her daughter Harriett in 1884, purchased the property. The Smiths continued rebuilding the farm and were active in their community. They established and helped to construct Zion Hill Baptist Church on the farm. They also donated land for the nearby Mt. Pisgah Methodist Church. Today, the owners of the Smith Farm and the residents of the community of

Built in 1858, the main dwelling at Light House Farm is a typical of many farmhouses built mainly in the 1840s and 1850s. Called an "I-House," it features a central hall with two rooms over two, chimneys at either end, and a two-story porch or portico.

Hawkinsville continue to honor the memory of Simon, a casualty of the war, and Isabella, who kept her family and farm intact in the most difficult of circumstances.

Harriett Hawkins and her husband, J. W. Smith, stand on the porch of her father and mother's pre-Civil War farmhouse. The family poses with a prized horse and colt in 1906.

While many families recount stories of foraging, the owner of the Gauldin Farm in Dyer County was assigned to procure food and supplies for the army as part of his military assignment. Of Irish descent, John William Gauldin enlisted in the 22nd Tennessee Infantry (CSA) at the onset of the conflict. According to Goodspeed's *History of Dyer County*, Gauldin was captured at Gallatin and taken to Louisville where he was imprisoned for five weeks. He was then transferred to Nashville and paroled after two weeks. He returned to his West Tennessee farm and his wife, Margaret, and their children. After two months at home, he joined the cavalry under Gen. Tyree H. Bell and served as provost until the spring of 1864. From then until the end of the war, he worked to secure provisions and supplies for the Confederate army.

Opposite Page: Sheaves of oats and other grains were a prize for foragers looking for feed for livestock..
From *Leslie's*

Ball's Farm

Crockett County

A farm purchased by David Nunn of North Carolina in 1845 was a large-scale antebellum agricultural operation in West Tennessee. Beginning with 240 acres just east of Maury City, Nunn added to his farm until it was well over 1,000 acres. Specializing in the production of swine, records reveal that Nunn sold over 10,000 pounds of pork in one year before the war changed the family's fortunes and lives.

The family recalls that Nunn "quit farming at the close of the Civil War because almost all of his livestock, farming tools and feed supplies had been taken from him by Union and Confederate troops." Elsey Nunn, his wife, died in 1872, and David followed three years later. Lucy Ann, the founders' daughter, had inherited 404.5 acres of the plantation in 1861. Her husband, Corday B. Revelle, served as a lieutenant and quartermaster in the 14th Tennessee Cavalry (USA) and died at the Battle of Fort Pillow, north of Memphis, on April 12, 1864.

In 1868, Lucy Ann married David E. Mayo, also a North Carolina native and the farm's overseer during her husband's absence and after his death. In the 1870s and 1880s, the Mayos concentrated on corn and cotton production and operated a cotton gin. Their efforts succeeded in weathering the hard times and preserving the farm for the Nunn and Revelle descendants who have operated the acreage as Ball's Farm since 1942.

At the Battle of Ft. Pillow on April 12, 1864, casualties were high among white and black Federal troops when Confederates overran the fort. Corday Revelle, husband of Lucy Nunn Revelle, was among those killed.
From *Soldiers in Our Civil War*

Dillard Brooks Farm

Dyer County

William Calvin (W.C.) Brooks and his wife, Tabor, married in 1853 and purchased a farm in 1863. Brooks opposed the war and decided to remain on the farm to take care of his wife, their small children, and his property. The disadvantage to this decision was that whenever soldiers passed by the farm, he had to hide. The continuous interruptions of going for cover interfered with chores, so the couple devised a plan that allowed Brooks to work outdoors and remain safe. He was small of frame and about five feet, four inches tall, so he shaved his beard and dressed in Tabor's clothing. It was a common sight to see women working in the fields, and Brooks was never bothered while so disguised.

As the war dragged on, though, Brooks began to feel guilty about not being involved and enlisted in the 6th Tennessee Cavalry (USA), commanded by Col. Fielding Hurst. Brooks began serving with this unit in February 1864. This was the time of Hurst's notorious visit to Jackson, Tennessee, when he demanded payment of more than $5000, or else he would burn the town.

Back in Weakley County, Tabor, whose five children ranged in age from 5 months to 10 years, was left with the responsibilities of caring for the family and maintaining the farm. Her aunt, Susan Daugherty, lived with them to help with the children and share the farm work. In a letter dated February 1865, in which Brooks sent regards to his wife, to "Aunt Susan," and to his children, he wrote that there was a "great deal of talk of peace" and, though he hoped to be home by spring to help with planting, he could not be certain this would happen. He advised his wife, "If you haven't hired anybody you had better not try to plant any crop for you can't tend it."

Brooks, along with the rest of his regiment, was mustered out in July 1865 in Pulaski (see next page). The Civil War Veterans Census of 1890 lists W.C. Brooks living near Martin with the further information that he incurred bronchitis while in service. The family recalls that Tabor lost her eyesight soon after her husband's return. With six more children, the years following the war could not have been easy. G.D. Brooks eventually acquired 50 acres of his parents' property and passed along the homestead and family lore to his son, Dillard.

W. C. and Tabor Brooks survived the war but continued to face illness and difficulties.

To all whom it may Concern.

Know ye, That William C Brooks a Private of Captain John H Edwards Company, (D,) Sixth Regiment of Tenn Cavalry VOLUNTEERS who was enrolled on the First day of February one thousand eight hundred and Sixty four to serve Three years or during the war, is hereby **Discharged** from the service of the United States, this Twenty Sixth day of July, 1865, at Pulaski Tennessee by reason of Special order No 12 Hd Qrs M.T. Div of Tenn

(No objection to his being re=enlisted is known to exist.*)

Said William C Brooks was born in Henry Co in the State of Tennessee, is Twenty Eight years of age, Five feet Ten inches high, Fair complexion, Blue eyes, Light hair, and by occupation, when enrolled, a Farmer

Given at Pulaski Tenn this Twenty Sixth day of July 1865.

* This sentence will be erased should there be anything in the conduct or physical condition of the soldier rendering him *unfit* for the Army.

[A. G. O. No. 99.]

John H Edwards
Captain
Capt. Commanding the Reg't.

Francis Jackson
Capt and A.C. M/o 6th
C.C. M.o D M/o

W. C. Brooks, who fought for the Union, was discharged in July 1865 in Pulaski, Tennessee, and made his way back to his family and farm in Weakley County (see previous page).

E.A. Cuff Farm
Benton County

The Tennessee River, winding through sections of both East and West Tennessee, was also a prized goal of both armies. On the western part of the river, in Benton County, is another farm that was kept in production during the war by women and children. Margaret Cuff of South Carolina founded the farm in 1847. She and her son, Francis Asbury Cuff, grew cotton and corn and raised cattle, sheep, and swine. The Cuffs were also well known for their high-quality sorghum molasses. Francis became the owner of the farm in 1861, but while serving in the Confederate army, he was captured and imprisoned.

His wife, Sarah Sykes Cuff, and their children took over the management and work of the farm for the long years that he was away. When Cuff returned after the war, he became an active member of the Grange, was a trustee of Flatwoods Methodist Church, and also did blacksmith work for the surrounding farms, making "plows, hoes, and other tools." The Cuff family also "operated a sorghum mill, establishing it as a major Benton County crop" and a source of postwar income for the county's farms and businesses for many years.

Mules turning a sorghum mill was a common sight in Benton County for decades.
Courtesy Tennessee Department of Agriculture, Oscar Farris Museum

Rhodes Farm
Decatur County

Decatur County, also on the western side of the Tennessee River, is the location of a farm established by John Prior Hill and his wife, Rebecca Aldridge, in 1853. The Hills worked the land until their daughter, Fannie Elizabeth, and her husband, James Johnson, acquired the property in 1860. For several months of the war, the couple continued farming, but Johnson enlisted in the 2nd West Tennessee Cavalry (USA) at Trenton in September 1862. Like many soldiers, he was a casualty of disease that sometimes decimated the army camps. In October 1863, Johnson contracted smallpox and was sent home, but he died while in route at Bethel Station in McNairy County, on November 13, 1863. His remains were returned to his farm, where he was buried in the family cemetery.

Fannie Elizabeth and her two young children, William Campbell and Sarah, lived in a log residence and grew vegetables and raised swine and cattle on their 140 acres. After the siblings inherited the farm, William purchased Sarah's half of the property. He also acquired other land over the years, but always held on to the original acreage that his mother had maintained during the difficult years following his father's death. Buried along with Fannie Elizabeth, who died in 1891, and James in the Hill-Johnson Cemetery are John M. Quinn and his wife, Mary Elizabeth Hill, sister of Fannie Johnson. Quinn, so the family story goes, first enlisted in the Confederate army. He was taken prisoner and then joined the Union army. After the Civil War, he fought in the Indian wars in the Western states, but returned to Decatur County where he died in 1903.

James Johnson died of smallpox while in route to his home in Decatur County.
Courtesy Bill Smith

Claudine Rhodes, and her husband, Cleo, maintain the Hill-Johnson Cemetery on the farm where her family has lived for generations.
Courtesy Bill Smith

Alexander Springs Farm
Lawrence County

At the southern end of Middle Tennessee, near the Alabama state line and on the Buffalo River, the Alexander family lost much of their harvest, as well as their stagecoach business, to Civil War combat and to foragers. In 1853, Absalom and Ellen Fields Alexander established an operation of 1,015 acres ten miles north of Lawrenceburg near the Maury County line. Their land was bordered on the south by the river and was adjacent to the Military Road, a major antebellum transportation route from Nashville to New Orleans completed under the supervision of Gen. Andrew Jackson in 1820. The family operated the McMillan-Alexander stagecoach stand on this road from about 1830 to 1862.

Although an ideal location for marketing and transporting agricultural products, the Alexanders realized that military activity along this major artery would bring disruption and destruction not only to the stagecoach business but to their farm as well. Furthermore, the north-south telegraph lines that ran through the farm guaranteed constant attention from combatants. Hoping to at least forestall some of the difficulties, the family dismantled their stagecoach stand and several outbuildings that stood along the Military Road and moved them three miles to the east.

The farm escaped the worst ravages of war until 1864, when Confederate Gen. Nathan B. Forrest's September raid stripped the property "of all food, crops and livestock," leaving the owner with only "one crippled horse." The family attempted to feed cavalrymen who were so hungry that they ate raw corn dipped in hog slop for seasoning. Two months later, a skirmish between Gen. John B. Hood's advance guard and Union horsemen took place on the farm. The busy thoroughfare that had brought prosperity during normal years brought near-ruin to the Alexander family and their land.

After the war, the family's landholdings were diminished. Mack Keller Alexander, son of the founding couple, became the owner of 369 acres in 1874 and was instrumental in rebuilding the area by financially supporting several schools and churches. The family built a home near the site of the old stagecoach stand. The first railroad in the county passed by the farm in 1883, and a paved thoroughfare (U. S. Hwy. 43) constructed in the early twentieth century followed the route of the Old Military Road.

Sanders Spring Forest Farm
Dickson County

On Jones Creek, one of the main waterways running through Dickson County, John and Susan West Sanders lived on a farm established by her family as early as 1808. Here they built a cabin that, with a late nineteenth-century addition, remains part of the homeplace. John died in 1848 and Susan lived in the house and "ran the farm until her death in 1876." In the agricultural census of 1860, her livestock was valued at $1312, and she harvested 785 bushels of corn. The family owned slaves and the site of some of their cabins is known to be across the now-paved county road from the house. During the Civil War, three of Susan's sons were in the Confederate army. Henry was captured and died from malnutrition in Camp Douglas, Illinois. John J. was severely injured in the Battle of Franklin but survived. Along with his brother, Thomas Berry, he returned to farm the land their mother had successfully operated before the war and had kept in production as much as possible during the conflict. In 1870, the value of the livestock was down to $600, which included four horses and four cows, and the family produced only 300 bushels of corn. The acreage remained intact, however, and the family slowly transitioned into the realities of postwar farming.

The pre-Civil War log house, with late nineteenth-century addition, is the earliest dwelling on the Sanders family farm.

Skelley Farm

Williamson County

James Crawford and Sarah Louise Potts Skelley of Williamson County were married in 1856 and lived in a log cabin on the "headwaters of Leipers Creek" near the Boston community. In 1860, the young couple had a five-year-old son and a daughter who had been born the previous year. For the Skelleys, the year 1862 changed their lives immeasurably. James enlisted in January, was badly wounded in the knee at Perryville, Kentucky, in October, was captured at Harrodsburg, Kentucky, and exchanged at Vicksburg, Mississippi in December. Disabled and unfit for service as an infantryman, he later had to have his injured leg amputated "above the knee by recommendations of doctor and authority of his officers." Family members recalled that he had a wooden peg leg, and, according to a statement by a neighbor, Henry Cook, he was never able to walk again without the aid of a crutch.

During her husband's absence and his years of disability until his death in 1895, Sarah Louise Skelley, like so many other women in her situation, had little choice but to work hard each day, tend the livestock and raise, cook, and preserve food. A weaver and seamstress, she dyed and spun wool from the farm's sheep to make clothing and bedding for her growing family. She gave birth to six more children between 1861 and 1871, including twins in 1867.

In 1913, Sarah Louise Skelley filed an application for a widow's pension from the state of Tennessee. At the time, she was living with her youngest son, Henry Potts Skelley, born in 1871, his wife, Mary Beasley, and their six children on their farm on another section of Leiper's Creek. Preston Ingram, the current owner of the Sparkman-Skelley Farm, had the house and farm listed in the National Register of Historic Places and, in 2001, placed the property in a conservation easement with the Land Trust for Tennessee.

Sarah Louise and James Crawford Skelley are buried in Cave Spring Cemetery.

Holly Hill was home to three generations of the Skelley family, including matriarch Sarah Louise, during the early years of the twentieth century.

Allendale Farm
Montgomery County

Allendale Farm was founded on July 11, 1796, just weeks after Tennessee became the sixteenth state on June 1. The rich land, much of which lies in the horseshoe bend of the Big West Fork Creek, successfully supported sheep, wheat, tobacco, and cattle – "the four pillars of income" until the Civil War. The farm's location near Clarksville in northern Middle Tennessee and the border state of Kentucky meant that the family was often visited by soldiers, usually Federals, who occupied the county seat after the fall of Fort Donelson in February 1862.

The staunchly pro-Confederate Bailey F. Allen, Sr., and his wife, Mary Jane Osburn, operated the farm through the Civil War and Reconstruction periods. The current owner of Allendale is their grandson, William B. Allen, Sr., whose own father, Bailey F. Allen, Jr., was born on the farm in 1863. The Allen family has preserved many documents, letters, and oral traditions from those troubled times.

On one occasion, a group of Confederate soldiers arrived at the farm. As William B. Allen, Sr., retold the story, "Snow was on the ground and they were cold," and the small house could not hold them all. "Grandpa Allen showed them a rail fence and told them they could use it for firewood." They lit the leaves that had blown up against the fence and the rails began to burn. They had a "long fire that all could warm themselves by for quite a while."

Bailey and Mary Jane Allen's children also had encounters with soldiers. Ella Allen, who lived from 1848 to 1930, was a "very high spirited young lady." One day, she headed to Clarksville, occupied and controlled by Federal forces. As she approached the picket post, the guards refused to let her pass. However, the fourteen-year-old, who was an excellent horsewoman, "just put the spurs to her horse and went on through the lines."

Her brother, thirteen-year-old Fountain Pitts Allen, tried to enlist

In 1896, members of the Allen family, along with tenants, gathered at the farm.

The 1858 brick house and a ca. 1800 log house are home to the three generations of Allens.

in the Confederate army, but when the authorities realized how young he was, they sent him back home. On several occasions, "Fount" visited the nearby Federal camp and boldly asked for gunpowder and shot, as well as permission to hunt game, so he could help feed his family.

William B. Allen, Sr., also recalls his father telling him of an incident that occurred just after the end of the war when "scalawags" and "ruffians" roamed the countryside. "Grandfather Bailey Allen had sold his tobacco, come home, and sat counting his money at the table." He heard a commotion at the back door and, fearing he was about to be robbed, put the money back in the sack and hid it behind some furniture. Masked gunmen burst into the house and told him they knew he had money. They threatened to whip him and worse if he didn't tell them where it was hidden. Aunt Betsy, an elderly spinster who lived in the house, heard it all and quietly got her pistol and came downstairs. With gun in hand she walked up to one of the men and said, "I believe I know who you are," and whipped off his mask. The man was a "ne'er-do-well" neighbor who had probably told the others about the money. The men left with nothing, and no one was harmed.

Throughout the war, foraging parties confiscated horses and wagonloads of corn and hay, as well as other provisions. As a result, the family had little to eat because so much of their food and livestock had been taken. "Allendale was all but dismantled" as a result of the war, explains William B. Allen, Sr. "Of the original 1,200 acres, all but 300 were lost to debt and federal seizure," advises Allen, but his family, with the help of tenants, returned to the tried-and-true formula of raising wheat, sheep, tobacco, and cattle on the reduced acreage. Bailey F. Allen, Jr., became the owner after his father's death in 1880, and his son, William B. Allen, Sr., took over in 1943. Since the Civil War, each generation has fought, in its own way, against varied challenges to preserve the farming tradition at Allendale. Allendale Farm is listed in the National Register of Historic Places.

SALTPETER AND GUNPOWDER MANUFACTURING

Saltpeter, essential for making gunpowder because of its high nitrate content, was vital to the war effort. When the Civil War began, the Confederacy realized that it did not have enough gunpowder, and states with large saltpeter reserves immediately geared up for production. Many areas of Tennessee, which has numerous limestone caves that contain saltpeter, were mined before and during the Civil War. Tennessee Gov. Isham G. Harris appointed a Military Board before the state seceded on August 24, 1861, and one of its duties was to encourage the production of saltpeter and gunpowder.

Despite crisscrossing the state, the Military Board initially had difficulty obtaining enough saltpeter for the gunpowder mills to remain in full production. The Board spent much of 1861 trying to secure enough saltpeter for the powder mills and soon turned to purchasing saltpeter from Arkansas and Alabama. By fall 1861, saltpeter production had increased, but the Confederate defeats in Kentucky and Tennessee, the fall of Nashville, and subsequent Federal occupation combined to disrupt saltpeter and gunpowder production in the state by late 1862.

Saltpeter caves were located on or near several Century Farms, and families recount various stories of saltpeter production. The Ward Farm, also known as Goddard Mountain Farm I in Van Buren County, has at least 24 identified saltpeter caves. The farm is near Big Bone Cave, which held one of the largest saltpeter mining operations in the state during the Civil War. Family lore tells of a slave named Ben who was hidden from Federal troops in the cave so he could continue to mine saltpeter. Grassy Cove (Cumberland County) also had saltpeter mines, and gunpowder was produced in "Powder Mill Cave" by families whose descendants still farm in the cove. The 2006 United States Survey of Saltpeter Caves, by Douglas Plemmons, lists 344 Tennessee caves by county with information about their dates of mining operation. Other Tennessee saltpeter caves mined during the Civil War included Nickajack Cave (Marion County), Pigeon Forge (Sevier County), Lookout Mountain Cave (Hamilton County), and the "Lost Sea" near Sweetwater (Monroe County).

Limestone caves, found in many parts of Tennessee including the hills surrounding Grassy Cove, were mined for saltpeter.

John C. Kemmer Farm
Cumberland County

Grassy Cove, about ten miles from Crossville, is a distinctive geographical area and prosperous farming community that was settled in the Cumberland Mountains in 1801. The stories of several of the farm families, most of whom supported the Confederacy, have been recounted in publications and by word of mouth through the generations. Not only was it a productive farming region, but caves in the area were also mined for saltpeter to make gunpowder. Because of the abundant livestock and food stores, soldiers often foraged in Grassy Cove. They also took any decent horses, leaving their broken-down mounts in exchange. To make it easier on these older and weaker animals, and perhaps to get the seed into the ground unobserved, the cove's farmers often plowed by moonlight.

The Kemmer name is a familiar one in Grassy Cove, and it is from land owned and settled by Andrew Kemmer that at least four current Century Farms have evolved, including the John C. Kemmer Farm. During the war years, Kemmer found it difficult to keep enough food for his family and the remaining stock when foragers roamed the countryside. He devised and built a false wall on one side of his living room to create a hidden vault. There he concealed and stored corn, rationing it to feed his family and their few animals until the next harvest, if they could save enough seed to plant.

In the course of the hard times, a neighbor, John Cox, came begging for some food to feed his hungry family. Kemmer asked the man to wait outside while he got some corn to share. The man's primary intention, apparently, was to see the location of the hiding place. The next day, about a dozen Union soldiers came to the farmhouse, tore out the false wall, and took all of the corn. In spite of this loss, Kemmer and his wife, Elizabeth Crook, who were the parents of eleven children, managed the keep the farm and their family together.

Grassy Cove has several Century Farms which trace their history to the early 1800s when their ancestors settled this community.

Alexander Farm

Loudon County

The Alexander Farm, founded in 1814 in what was then Roane County, is situated on an early pioneer route, the Hotchkiss Valley Road, that wound between Knoxville and southeast Tennessee. In 1860, the descendants of James and Mary Alexander were living in a two-story log house built by the founding couple. Outbuildings held quantities of wheat and corn, and livestock roamed the acreage. William Lawson Alexander, the third-generation owner, served in the Confederate army while his wife and at least two other family members lived on the farm during the war years. In mid-November of 1863, the farm was in the line of both the Confederate and Union armies as soldiers marched through the valley on their way to Knoxville after the Battle of Chickamauga. Union veteran William Todd of Albany, New York, wrote about this campaign in his postwar memoir, and he devoted a full page of his account to a foraging incident that he participated in at the Alexander Farm.

Moving northeast from Chickamauga, Todd and his company were stationed near the town of Loudon and were "obliged to scour the country in all directions" for grain and food for their animals. Todd recalled that they came to a house occupied by a "Scotch family named Alexander," where two women, whose husbands were serving in the Confederate army, lived with an elderly man. When the Alexanders told the soldiers that "they had nothing in the way of corn or fodder for us 'Yankees,'" Todd and his comrades, who had heard from an informant that the

Though not lived in for over 100 years, the Alexander house remains intact and in use on the farm.

Remnants of the Hotchkiss Valley Road are still well-defined on the Alexander Farm. This early trace brought troops within a few yards of the family's dwelling and barns.

farm did have provisions, proceeded to search the farm. In one barn, hidden under a pile of corn stalks, they found over one hundred bushels of corn still on the cob. They loaded about half of the corn and two wagonloads of the cornstalks to take back to camp.

As they had been ordered, the Union soldiers offered a receipt to the family that entitled them to redeem it later for market value "provided they could prove that they had been good Union people during the war and had never rendered 'aid or encouragement' to the enemies of the Union." In 1886 when Todd wrote his memoirs, he still vividly recalled the "looks of scorn and contempt" of the elder woman as she received the paper, tore it in pieces, and "stamped upon the fragments in the most approved and dramatic manner." She looked at the Federal soldiers as "though she would have enjoyed treating us in the same way." Todd described his return to camp after their day's work to find that half-rations had been issued. He wrote that as "our haversacks contained two or three days' supply of biscuit, and in addition we had a number of chickens and young pigs 'purchased' during our expeditions, short commons did not trouble the foraging party."

The log house lived in by the family during the Civil War and afterward, remains on the farm, and the Alexander family continues to own and work the farm. The property became part of Loudon County when it was created in 1870.

Jenny-Ben Farm
Greene County

The Jenny-Ben Farm originated with the Farnsworth family, who crossed the mountains from Virginia to what was then North Carolina as early as 1787. John W. Farnsworth and his wife, Elizabeth Parman, lived on 257 acres, grazing cattle and oxen and raising wheat, corn, vegetables, and fruit for their table. The Farnsworths' ten-room frame house, built ca.1840, was large enough to accommodate their growing family. By 1862, they had five young children. John enlisted in the Confederate army, was captured, and finally sent to Camp Douglas in Chicago, Illinois. Camp Douglas was named for Sen. Stephen A. Douglas who opposed Abraham Lincoln in the presidential election of 1860. Camp Douglas was somewhat comparable to Camp Sumter in Andersonville, Georgia, where Union prisoners faced inhumane conditions. The Illinois winter of 1864 was particularly brutal for men without adequate clothing and food, and John Farnsworth died there in November.

Farnsworth's family learned that he had died of starvation and that, although he had first been buried at the prison, his remains were reinterred in a common grave in Oak Woods Cemetery, about five miles from Camp Douglas. This graveyard may be the largest burying ground for Confederate soldiers in the North. A granite monument,

This sketch, part of John M. Copley's reminiscences of Camp Douglas, was published in 1893.
Courtesy *Documenting the American South*, The University of North Carolina at Chapel Hill Libraries

PRISONERS STRIPPED AND SEARCHED IN THE SNOW AT CAMP DOUGLAS.

erected in 1893 by several camps of the United Confederate Veterans and local citizens in the cemetery's "Confederate Mound" section, is inscribed as follows:

ERECTED
TO THE MEMORY OF
SIX THOUSAND SOUTHERN SOLDIERS
HERE BURIED WHO DIED IN
CAMP DOUGLAS PRISON
1862-5

Back in Greene County, which was largely Unionist, some neighbors took advantage of Elizabeth Farnsworth's situation and robbed her. Federal troops also foraged on the farm, found meat and other supplies she had hidden, and "took it all." The story that has been passed down through the years is that "feather beds were cut open in the yard so the feathers would blow away." Family tradition holds that the soldiers even sat Elizabeth "on the stairs and took her shoes."

Farnsworth and her children escaped physical harm, however, and managed to stay together and survive. With little help, she raised all five children to maturity. Because money was so scarce after the war, she sold part of her land to pay taxes. Her son, Benjamin, became the owner of the farm in 1878 when he was 22. It is his line, now represented by his great-great-granddaughter and her family, that passes along the harrowing wartime stories of Elizabeth Farnsworth and her children.

Elizabeth Farnsworth died in 1913 and is buried in Soloman Lutheran Cemetery in Greene County.

Tilson Farm
Unicoi County

In the mountains of East Tennessee, Elizabeth Beals Tilson and her three young children, George, Marion, and Catherine, all under ten years of age, looked forward to the letters from husband and father, James W. Tilson, who was in the Confederate army. Their home was an 18' x 20' log house with a loft, that had been built before James purchased the land in 1856. As Elizabeth cared for her children and managed the farm, she received at least seven letters from James before he was killed at the Battle of Chickamauga on September 20, 1863. Nearly two years later, however, her brother-in-law, Capt. William E. Tilson, received a letter from Jacob Ottinger dated April 14, 1865. In it he explained that he had just learned that a letter he had written five days after the death of James Tilson had never reached Elizabeth and her children. Ottinger offered a description of the injury, final hours, and death of James Tilson. He told of burying him with his blanket "roped around him good" and of how he had put his knapsack "in the bottom of the grave for his head to lay on." Ottinger also said he had carved a wooden plank with Tilson's name on it should anyone try to locate his burying place. With his last breath, James had asked Ottinger to write to his wife and say to her "not to lament for him" and that "he was going to a happier home and that he was better off than those that was left behind."

Elizabeth Tilson, who was thirty-three years old when James died, remained a widow for the rest of her long life, raised her three children, and worked her farm which became part of Unicoi County when it was formed in 1875. She cultivated corn, tobacco, and wheat, primarily, and raised cattle on 132 acres. She was eighty-two when she died in 1914. Her sons, George and Marion, were lifelong bachelors who lived and worked on the farm where they were born until they died, a day apart, in 1936.

The 1850s log dwelling is reached by the old farm road which passes by several historic outbuildings.

Daughter Catherine Tilson and her husband, William Mashburn, built their own house on the farm and lived there with their six children. Janice Rhodes, who is the current owner along with her husband, Leon, is the great-great-granddaughter of the founders. She and two of her three brothers were born in the same log house where Elizabeth and her children lived through the long and difficult years of the Civil War and afterwards. The Tilson Farm is listed in the National Register of Historic Places.

Directly across from 1850s log house is this cabin that was constructed in the 1870s. Behind it is the spring house.

A few farmers were fortunate to receive pay, at least some of the time, for products that they supplied to the armies. In 1862, Benjamin Franklin Earnest inherited **Elmwood Farm**, *established in 1777 in* **Greene County** *on the Nolichucky River. He realized a profit as he sold flour to Union and Confederate troops when they sought supplies in East Tennessee. This rare stone and log blockhouse at Elmwood was built ca. 1785 and is listed, along with the farm, in the National Register of Historic Places.*

Michael Krouse Farm
Washington County

The history of the Michael Krouse Farm in upper East Tennessee includes a vivid recollection by Isabell Krouse Sherfy of how her father, Daniel Krouse, stored wheat between the joists of their house for safekeeping from foragers. A hole at the bottom of the hidden space allowed the family to remove only what they needed. Their home was a stopping place for Union guide Capt. Daniel Ellis, also known as the "old Red Fox," as he made his way to and from Carter County piloting men, slaves, and fugitives across the mountains to join Union forces in Kentucky. Daniel Krouse and his wife, Susannah Wine Krouse, were the parents of nine children who were born before, during, and just after 1865.

David Sherfy was with the 11th Illinois Infantry and saw action in many battles, including Vicksburg, Shiloh, and Franklin. After the war, he first went to Centralia, Illinois, and worked in a haberdashery. He returned to Washington County to learn that his stepfather had died during the war and that his mother

Unionist David Sherfy and his second wife, Isabell Krouse Sherfy, represent two families who helped to rebuild their community after the war (see David Sherfy in his Civil War uniform in "Photography" inset).

and sister, penniless and with no way to support themselves, were living on his uncle's farm in "dire circumstances." He immediately took care of their needs, and then began to purchase property and establish himself. He went into the nursery business, supplying fruit trees to the farms whose orchards had been devastated during the war years. After the death of his first wife, Mary, he married his neighbor, Isabell Krouse. In addition to the Michael Krouse Farm, surrounding farms connected by family and community history during the Civil War and Reconstruction are Pioneer Homestead Farm and Lone Pine Farm. A collection of documents, objects, and artifacts, which illustrates the lives of these families as they farmed before, during, and after the war, is displayed at the Knob Creek Museum in Johnson City. A copy of a diary kept by Anna Krouse from January 1858 to April 1865, as well as quilts, coverlets, and other items made by Susannah Krouse and her daughters, are part of the varied collection of the granddaughter of Isabell and David Sherfy, Margaret S. Holley, and her husband, George Holley. Mrs. Holley's father was born when David Sherfy was 50 years old, and her own father, John Sherfy, was 40 when she was born. She is proud to be among a select few Tennesseans living today whose grandfather was a Civil War veteran.

Susannah Wine Krouse and her daughters regularly pieced and worked quilts like this one which remains with her descendant, Margaret S. Holley.

END NOTES

Corona Farm

Joanne Cullom Moore, "The Devil's Elbow." *West Tennessee Historical Society Papers, XXXV* (1981):5-24.

Skelley Farm

Besser, Susan. "Sparkman-Skelley Farm, Holly Hill Farm, Williamson County, Tennessee." National Register of Historic Places Nomination Form, Tennessee Historical Commission. Listed 2000.

Tennessee Confederate Pension Application, # 4719, Tennessee State Library and Archives, Nashville.

In 1891, Tennessee established a Board of Pension Examiners to determine if Confederate veterans or their widows were eligible to receive a pension. The first of the widows' pensions were paid beginning in 1905. These applications contain family history as well as service records.

Allendale Farm

West, Carroll Van, Elizabeth Humphreys, Jessica Bandel, Jessica French, and Amy Kostine. "Allendale Farm, Boundary Increase, Montgomery County, Tennessee." National Register of Historic Places Nomination Form, Tennessee Historical Commission. Listed 2013.

Alexander Farm

Todd, William. *The Seventy-Ninth Highlanders: New York Volunteers in the War of the Rebellion*, 1861-1865 (Albany, NY: Brandow, Barton and Co., 1886), 347.

Kemmer Farm

Harvey, Stella Mowbray, comp. *Tales of the Civil War Era* (Crossville, TN: Cumberland County Civil War Centennial Committee, 1963).

Michael Krouse Farm

George W. Holley. *David Preston Sherfy, Kinsman and Cavalryman* (Knoxville, TN: Earthtide Publications,1986).

Refer to the "Bibliographical Essay" for additional resource materials.

Freedom to Farm

Freedom to farm their own piece of land was a dream for many enslaved men and women. With a plot of earth, they could use their labor, skills, and abilities to provide for their families and then leave some property to their children and grandchildren. Along with a farm, communities with schools, churches, businesses, and cemeteries were central to the life and liberty that enslaved Tennesseans imagined.

That dream began to take shape when President Abraham Lincoln announced the Emancipation Proclamation in September 1862, declaring that, after January 1, 1863, all slaves in the areas of rebellion would be free. Although the Proclamation excluded Tennessee since most of the state was under Union occupation, it nevertheless encouraged many enslaved people to leave their owners and seek freedom. On April 7, 1865, Tennessee ratified the Thirteenth Amendment, which became part of the U. S. Constitution in December 1865, forever abolishing slavery in the United States. Despite this achievement, the prospect of owning property was still elusive for most former slaves. It would be years before most freedmen or their children could acquire their own land.

Still, former slaves were now able look for the best situation they could procure for themselves and their families. For various reasons, though often as an incentive to keep the ex-slaves nearby and working on the farm, some former masters gave acreage to African American or offered it to them for a very reasonable price. Many blacks found sharecropping or tenant farming the best option, though often a disappointing one. When all was taken into account, this arrangement of working for former masters or other landowners on shares or for a percentage of the harvest, sometimes under yearly contracts, prevented former slaves from accumulating enough money to purchase land, much less highly productive land. Other newly-freed men and women entered into binding labor contracts with white landowners to do farm work for wages, housing, clothing, seed, and food. Those men and women who had specific and marketable skills and tools, and who worked independently and lived frugally, often achieved land ownership in their lifetime. For many families, however, it was the sons and daughters of the ex-slaves who were finally able to purchase their own farms.

Neighborhoods and communities of former slaves appeared as they purchased land and established commercial ventures. Some communities grew up around former contraband camps, where black refugees had come for protection and work, near Federal encampments. The Cemetery community in Rutherford County, on land adjacent to Stones River National Cemetery, is one example of a place where black men, women, and children stayed after the war's end to begin a new life with

(Opposite page) By 1880, the Jerry and Mae Jane Robertson family of Dickson County was living in Promise Land, a community of freed blacks who had formerly lived and worked as slaves at nearby Cumberland Furnace. Many of the residents continued to labor at the iron works, but now also farmed their own land. Their descendants remain in Promise Land.
Courtesy *Promise Land Historical Site*

their land, homes, a school, churches, and burying ground.

Other communities, like Promise Land in Dickson County, were carved out of existing farmland by families who combined their resources (in this case, war pensions) to purchase property where former slaves could live. Individuals and families who had worked on farms and at the iron works at nearby Cumberland Furnace established a vibrant neighborhood whose story continues today. This pattern occurred in many counties, and remnants of those Reconstruction communities remain as significant evidence of African American heritage.

The Black Family Land Trust reports that African Americans amassed 15 million acres in the South between 1865 and 1919. From that point on, however, the number of black farmers declined steadily throughout the twentieth century. Jim Crow laws and a lack of financial support account for the large numbers of black farmers who left farming to pursue other occupations. Today, farmers of African American descent number less than 18,000 nationwide, and they own less than one percent of all farmland in the United States. Within the Tennessee Century Farm program, only eight certified farms are owned by the descendants of African American founders. These stories, along with other Century Farms whose stories are especially closely connected to the history of African Americans, are among the most remarkable told by Century Farmers. Each story is tangible legacy of the time when enslaved peoples looked to a future that included the freedom to farm.

Rising Sun Farm
Wilson County

The Reverend Andrew Hunter Davis was 86 years old in 1897. The former slave and overseer on the Rising Sun Farm was given land in 1864, on which he built a church where he preached for many years.

When James Harvey Davis died in 1864, his daughter, Emma, gave land to the farm's black overseer, Andrew Hunter Davis.

The history of the white and black Davis families who lived and worked on the Rising Sun Farm continues to engage current generations. James Harvey Davis, was among the men who surveyed land along the Cumberland River in the late 1700s. In 1824, he purchased 98 acres which became the core of a large plantation. Davis was married first to Penelope Drake and after her death married Eliza Thomas; each wife gave birth to eight children. Davis expanded his landholdings until he owned about 2000 acres in Wilson County, around the Laguardo community, and a plantation in Mississippi. Davis and his family relied on about 100 slaves to work the farms. In 1860, when James H. Davis was in his 70s, he began work on an impressive two-story brick house which was built largely by his slaves. The house was not completed by 1861, but was finished to the point that Union soldiers occupied it and used the top floor as a lookout. Two of the Davis sons, John and Bunkum, joined the 18th Infantry (CSA) and were captured at Fort Donelson before being exchanged to continue their soldiering. Bunkum became the chaplain of the 18th Infantry for the remainder of the war.

In 1864, James Harvey Davis died, and his daughter, Emma, became the owner and manager of the family's land. Like her father, she relied on the farm's longtime overseer, Andrew Hunter Davis, a slave who had been bought by James Harvey Davis from the Hunter family of Hunter's Point, also in Wilson County. Not long after her father's death, Emma Davis gave Andrew Davis a piece of property. He took part of it and built a church which is still standing and known as Andrew Tabernacle C.M.E. Church. He became a preacher and held services in this church for many years. The Reverend Andrew Davis married twice, had a large family, and was well respected in the community. Since 1978, descendants of Andrew Davis have travelled from many states to gather at the church every two years. They also visit Rising Sun Farm where Alfred Adams, the great-great grandson of James Harvey Davis, lives. In the cemeteries are ancestors of both Davis families whose history is bound to the surrounding land.

Matt Gardner Homestead
Giles County

In southern Giles County, near the Alabama state line and just outside the small town of Elkton, is the Matt Gardner Homestead. Within the traditional African American community of Dixontown, on the banks of the Elk River, the farm is named for a man whose influence remains strong in the history and culture of the area. Matt Gardner was born into slavery in Chester, North Carolina, in about 1848, to Martin and Rachel Gardner. The Gardners, who held high status within the plantation household, were allowed to live together with their children and to acquire some money and small pieces of property.

At some point, the Gardner family came to Giles County where they soon were recognized as a leading family in the African American community. In 1870, the 22-year-old Matt Gardner married Henrietta Brown, and they began to build a new life together. Family history records that by the time he married, Gardner "had money to buy land and started off with three hundred acres." He soon gained respect and recognition as a minister, farmer, and businessman, and his reputation continued to grow in the succeeding years.

Unlike many of his fellow freedmen who acquired land in the years after emancipation, Gardner was able to maintain and expand his farm successfully during the following decades, even within the realities of "Jim Crow" segregation and increasing racial discrimination. By 1900, Gardner had emerged as a local leader, and his farm was the focal point of the area's African American community. He owned some 500 acres and built a new farm house in 1896.

He operated a "storehouse" for his neighbors where he sold merchandise on cash, credit, or barter. According to the family, Gardner loaned money to many of his black neighbors so they could purchase their own property. His account books tell the stories of other prosperous southern Tennessee freedmen who started a variety of businesses, including blacksmith shops, stores, and mills.

Gardner, also a preacher for a Primitive Baptist congregation, advocated for a public school for African Americans in Elkton. He provided room and board for teachers in his home. When the Rosenwald school-building program emerged in the early twentieth century, Gardner led the community's effort to match the contributions of the fund and the state. Named for philanthropist Julius Rosenwald, the program helped communities throughout the South build schools for African Americans. In time, a new four-room school was built within a half-mile of Gardner's residence. Gardner remained an active community leader, farmer, preacher, and businessman until his death in 1947.

Matt Gardner, a former slave, was a farmer, merchant, and preacher. He and his wife supported education for African Americans in Giles County.

Gardner diversified his crops and livestock which included a swine herd. In 1942, Gardner received a certificate for growing seventy-five percent of the food consumed by his family under the "Tennessee Home Food Supply Program."

The Matt Gardner Homestead, listed in the National Register of Historic Places, is owned and operated by his descendants and interprets the story of the Gardners and associated families, their way of life, their social, religious, agricultural, and educational contributions, and the African American heritage of Giles County.

McDonald Craig Farm
Perry County

The Tapp Craig Branch, off the North Fork of Lick Creek in Perry County, is named for the ex-slave who purchased land in 1871; the water flows through the farm where his descendants live today.

Established in 1819, Perry County is located on the Western Highland Rim and bordered by the Tennessee River. Many of the county's early settlers were slaveholders, and among them was Andrew Craig. One of Craig's male slaves, Tapp, took his owner's surname, as did many enslaved men and women.

Tapp Craig and his wife Amy, a slave of neighboring farmer Andrew Guthrie, were married and had four children before emancipation. The Craigs continued to labor for their respective owners throughout the war years and afterwards chose to work together on the Guthrie Farm. On Christmas Day in 1871, the Craigs gave a yoke of oxen and $150 cash as the down payment on 110 acres owned by local farmer Samuel Young. The balance of $400 was paid in full over the next two years, and the Craigs became the first African Americans to own land in the county.

Engaged in timber production (still one of the primary industries in the county), the Craig family selectively cut the trees, harvesting only the older specimens. From these, they produced railroad ties, barrel staves, and other products, as well as tree bark, which was gathered and transported to a tannery located at Mousetail Landing on the Tennessee River. They also participated in the burgeoning peanut business, a principal cash crop of the county for many years. As an easy-to-grow and lucrative commodity, peanuts helped the Craigs to maintain their farm throughout the nineteenth century and to prosper when others were unable to keep their property. The farm they established was the beginning of a community of African American farmers, which included their only surviving son, William, who purchased property nearby in the 1880s. William's grandson, McDonald Craig, owns and farms the original 1871 acreage which is listed in the National Register of Historic Places.

Tapp and Amy Guthrie Craig, daughters, Tenna and Mary Jane, and son William (not pictured) purchased a farm in Perry County six years after the Civil War.

Drake Farm
Sumner County

The story of the Drake Century Farm in Sumner County begins with George Bullock, who was born a slave in 1824 in Mt. Sterling (Montgomery County), Kentucky. Family tradition holds that George was the illegitimate son of his master, Nathan Divine. The Divines owned thirteen slaves in the 1850 census. The family's oral history recounts that when Susan Divine, the daughter of Nathan and his wife Sallie, married James Bullock in 1846, George was given to his white half-sister as a wedding present by their father. In 1854, the Bullocks moved to a farm of about 300 acres in Sumner County, Tennessee, and brought George with them. In 1860, the Bullocks owned several slaves including George who was then about thirty years old.

The 1870 census lists George Bullock and his wife, Maria, also a former slave born in Kentucky, living with their three children in Sumner County, but not owing property. George was a "farmer" while Maria is noted as "Keeping House." They may have been working as tenants or sharecropping on the Bullock Farm. In January 1876, however, George Bullock purchased from the Chancery Court in Gallatin 160 acres in the community called Macedonia, which was settled mainly by freed blacks following the Civil War. The family believes George was able to read and write, skills few slaves were permitted or able to acquire. They do know that George and Maria supported education, and their son, Henry Bullock, attended Wilberforce College, an institution founded in Ohio in 1856 for the purpose of educating African Americans. When Henry died, his daughter, Alice, and wife, Frances Thompson Bullock, survived. Alice Bullock was raised mainly by George Bullock for Maria had died in a house fire. Alice Bullock Drake inherited the farm on the death of her grandfather in 1916. Frances E. Drake Malone and her husband, Richard, have lived on and operated the farm for many years. Mrs. Malone has spent hours researching her family's history and comments that "the past is what it was and cannot be denied or changed, but I am blessed that life is different today."

This likeness of George Bullock, a former slave, is prized by his descendants who remain on the farm he established in 1876.

Robertson Farm
Hardeman County

The strong desire for at least a basic education for themselves and their children drove former slaves to establish schools in the decades after emancipation. In West Tennessee, formal education for African Americans can be traced to the contraband camps operated by Federal troops as they advanced through the lower Mississippi Valley. Slaves deserted the plantations that they had worked on to take up residence in these camps, like the one in Grand Junction in the southwestern part of the county.

During the war, Gen. Ulysses S. Grant appointed Chaplain John Eaton, Jr., to coordinate educational programs for ex-slaves in West Tennessee including reading, writing, and technical and domestic training for adults and youth. As the state superintendent of public schools from 1867 to 1870, Eaton continued to push for some reforms, and the Public School Law of 1867 required each civil district to establish a school for black children when the scholastic population exceeded 25 students. By 1873, 120 African American pupils

Many slaves had their first taste of freedom when they left farms during the war to travel to contraband camps where they lived and worked behind Federal lines.

Courtesy *Leslie's*

were enrolled in Hardeman County schools, and 10 African American adults had earned teaching certificates.

Crawford Robertson was born a slave in Arkansas in 1856. By the time he reached adulthood, the war was over, and he had moved to Tennessee where he was able to accumulate nearly 200 acres in Hardeman County by the latter part of the nineteenth century. "Ambitious, industrious and energetic," according to his grandson Evelyn C. Robertson, Jr., Crawford Robertson "demonstrated his skill as a farmer and a carpenter in an impeccable business-like manner." Though Robertson and his wife, Cora Pierce, lacked a formal education, they were determined that their seven children would have the opportunity to attend school. When local schools for blacks ended at the eighth grade, the Robertsons sent their children to Nashville and Memphis to complete their secondary education. Robertson's daughter, Myrtle, and son, Evelyn, Sr., graduated with degrees in Home Economics and Agriculture, respectively, from Tennessee Agricultural & Industrial State Normal School, now Tennessee State University.

In the 1920s, Crawford Robertson was among a group who established the Allen-White School in Whiteville. He was the treasurer of the organization that raised funds that dwarfed the contribution of the Julius Rosenwald Foundation. This Hardeman County organization was the most successful local fund-raising group in the state. From 1933 to 1960, the Allen-White School was the only high school for African Americans in Hardeman County. Myrtle, Crawford's daughter, helped her father canvass the community by horse and buggy for financial contributions for the school; she eventually taught at Allen-White School for forty years. Cora Robertson was involved for many years with educational efforts directed at farm wives through the county's home demonstration clubs. Evelyn C. Robertson, Jr., who owns the family farm today, continues the agricultural and educational traditions of his family.

Crawford Robertson, his wife Cora, and daughter Myrtle actively supported education for former slaves and their children.

The Robertson family of nine lived in this 1906 frame house which remains on the farm.

Butler Farm
Rutherford County

In the 1860 census of Rutherford County, the geographic center of the state, 196 free blacks were listed. Some were farmers, including Isaac Williams, who owned more than 1,250 acres, while Caroline McKnight held title to 125 acres. However, most of the county's African Americans were enslaved. A slave market was located on the north side of the town square, and many of the large farmers, including the Maneys of Oaklands plantation, relied on their slaves to raise large row crops and tend herds of livestock.

The Federal occupation of Murfreesboro, along with the construction of Fortress Rosecrans, attracted slaves to the town and surrounding countryside where they sought freedom, protection, and jobs within Union lines. Some ex-slaves lived in abandoned houses while others resided in contraband camps. As early as 1865, the Freedmen's Bureau reported that schools were operating in the county for black students; Murfreesboro was also the site of one of the bureau's hospitals.

By 1880, former slaves, Josiah Butler and his wife, Martha Lillard, along with their seven children, were living and working on a farm on the Old Woodbury Highway east of Murfreesboro. The family raised corn, cotton, and vegetables on the 26-acre Butler Farm. In 1889, Josiah Butler purchased land for a family cemetery.

The oldest Butler son, Perry, born into slavery in 1849, was the next person to own the farm. He married Alice Henderson, the daughter of Isaac and Lavinia Henderson, two former slaves of Rutherford County Judge Logan Henderson, who owned Farmington, a sizeable plantation. During their ownership of the Butler Farm, Perry and Alice, the parents of ten children, founded a school and a church, Butler's Chapel, on their property. The Butler family continues the tradition of agriculture, education, and service passed down from their ancestors, many of whom are buried in the family cemetery.

Perry Butler, born a slave in 1849, was the second owner of the family farm. He was buried in the Butler Family Cemetery in 1920.

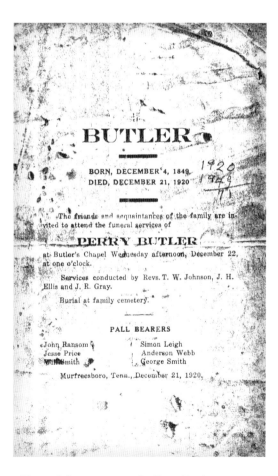

Funeral Announcement for Perry Butler

Tony Angus Farm
Rutherford County

Also in Rutherford County near the Bedford County line in the New Town community is the farm founded by ex-slave Jesse Landrum and his wife, Cora McClain, in 1891. Landrum was a skilled blacksmith and may have been the only African American smithy working in that area at the time. Because of his skills and industriousness, Landrum owned several parcels of land. The Landrums, including two children, Beulah and Genie, worked hard at being self-sufficient. They grew vegetables, had a milk cow, and also grew hay, corn, and cotton on 40 acres. Beulah became the next owner of the farm. Her husband, Charles Lanier, who played baseball in the Negro Leagues, and their children, Jessie, Mable, and Everlee, grew tobacco and cotton as cash crops. The family had a "kind of community park" on the farm which included a baseball diamond and picnic area. The fourth generation owner, Tony Scales is the son of Everlee. He and his wife, Maxine, returned to the farm after retiring from Ford Motor Company. They have built a new home and barns, and they run a herd of Black Angus on the land that was purchased by his great-grandparents.

Following the Civil War, Jesse and Cora Landrum worked for over 25 years to purchase their farm.

Nelson Bond's Oakview Farm
Haywood County

After the December crop was harvested in 1867, Nelson Bond, born in 1839, married Harriette Johnson, born in 1840. Like many freed men and women in the years following emancipation, the Bonds married, a privilege legally denied enslaved men and women. The couple farmed on rented land, or sharecropped, began a family, and saved as much money as possible. After twenty years of working on shares, Nelson and Harriette Bond purchased, in partnership with Dick Mann, 120 acres northwest of Brownsville from P.H Mann, a landowner and former slaveholder. When the Bonds completed making payments on the land, Dick Mann relinquished all rights to the farm in January 1891. At the time, Nelson and Harriette Bond and their children were active members of the Woodlawn Baptist Church congregation, one of the oldest African American churches in the county. After about 1900, they helped establish Oakview Baptist church, also a still viable congregation.

Nelson and Harriette's cash crop was cotton, but they also grew corn, sorghum cane, sweet potatoes, turnip greens, and peas to feed the family. Their livestock included mules, cows, and hogs, while chickens, and guineas rounded out the self-sustaining farm. Over the years, Nelson was able to purchase another eighteen acres; deeds indicate that several members of the Bond family also owned tracts of land in the area.

Nelson and Harriette had seven children, at least four of which lived on the Bond farm. After Harriette passed away, Nelson married Lucy Usher, a widow living in the area. When Nelson died in 1914, he bequeathed Lucy a dower and homestead while his surviving children inherited the remainder of the farm.

Luster Farm
Williamson County

Nelson Luster was born a slave in October 1834. His father was born in North Carolina and his mother was a Virginia native. In 1870, Nelson Luster and his wife, Betsey, were living in the 21st District of Williamson County with their four children, aged four months to eight years. They were likely tenants or sharecroppers.

Their second son, Grant Luster, Sr., born in May 1864, purchased a farm of just over 80 acres southeast of the county seat of Franklin in the Arno-College Grove area in November 1906 from Henry Graves. Luster and his first wife, Anna, born in May 1865, had four children, Grant Jr., Alex, Jennie, and Mattie. They were living on the acreage in 1910 where Luster's occupation was listed as "general farmer." After Anna's death, Grant Luster, Sr., married Sallie Jones in 1920. Nelson Luster, II, the son of Grant Luster, Jr., carries the name of his ancestor and owns the farm today. His son, Anthony W. Luster, observes that "Families of the past, while working with hand tools and beasts of burden seemed closer knit than perhaps today, but the legacy, the heritage, and the bond of the land still continues."

Grant Luster Jr., and his second wife, Mattie, acquired the family farm in 1931.

POSTWAR LABOR HOUSING

With the emancipation of slaves, newly freed men and women and their families had limited options to begin a new life. In many cases, farming or domestic skills were their only occupations. In the evolving post-Civil War agricultural system, farms and plantations that had relied on slave labor could not rebuild and return to production without help. In this environment, a labor system emerged that allowed former slaves to live on a farm and either work for the owner for wages (tenancy), or for a portion of crops produced on the land in return for their labor (sharecropping). As intended, both landowner and tenant or sharecropper would benefit from this system. Farms could return to production, and freed men, women, and children had a place to live while earning food, supplies and money.

This system did work for some black farmers, who slowly began to save money and accumulate their own possessions in hopes of one day buying their own land. For others, however, this situation was little better than slavery by another name. While arrangements differed from farm to farm, many tenants paid rent for housing—either former slave cabins, refurbished outbuildings, or newly-constructed tenant housing. The workers were charged by the landowner for food, seed, and tools. After all was tallied, workers often found themselves in debt year after year. Many employers took advantage of this system and would not allow tenants to leave until their debt was paid in full. In these situations, the debt owed by the workers could last for years, or even generations.

While the majority of tenants were black, hard economic times after the Civil War resulted in whites also taking up tenancy or farming on shares. With the migration of many Southerners to jobs and opportunities in the North, and the increased mechanization of farming in the early to mid-twentieth century, tenant farming generally diminished in the South by World War II. While increasingly rare on the twenty-first-century landscape, a few tenant houses remain as evidence of the dominant agricultural system that emerged after the Civil War.

*This house was built on the **James Farm** in what is now **Loudon County** between 1865 and 1870. It was lived in by former slaves and their descendants until about 1940, and is typical of the housing provided to sharecroppers and tenant farmers following the Civil War.*

Sugartree Farm
Robertson County

Sugartree Farm, which is owned by the Alford family, provides a different perspective on the postwar ownership of farms by African Americans. In Robertson County, Wessyngton Plantation was founded by Joseph Washington, a cousin to President George Washington, after he migrated to Tennessee in 1796. The farm he established is one of the most famous plantations in the South. At one time, Wessyngton, which is the Saxon spelling of Washington, was the largest producer of dark-fired tobacco in the state and perhaps in the world.

The butler and house manager of the 1819 brick dwelling at Wessyngton were Lawson and Marina Washington who, like so many people in bondage, took their owner's surname. Following the Civil War, the couple and their son, Foster, purchased 100 acres adjacent to the plantation where they had been enslaved. Here they farmed and operated a large orchard. In October 1884, Marina died and was buried on the farm. Of at least twenty graves in the cemetery, hers is one of only two with an engraved headstone. For unknown reasons, just two months later, on December 23, Lawson and Foster sold their land to Ben L. Alford, a white farmer. The Alford family has worked the land since that time and live on Sugartree Farm today. The graveyard marks the passage of the Lawson and Marina Washington family from enslaved workers to free farmers. The late Reverend Dr. Ben R. Alford wrote that the story of the former slaves "always conveys a sense of pathos and respect for these early owners as our family looks back to the origin of our association with this land." He further noted that this is a "place that holds great joy for us, but we know it was once home to both Native Americans, and African Americans. The land has been watered by rain, sweat, and tears."

On the marker of Marina Washington, among the first former slaves to own property in Tennessee, is this epitaph:

*"Notice grandchildren as you pass by
I am waiting in glory for them"*

Numbers of former slaves served in the United States Colored Troops. After the war, many returned to their families to farm as sharecroppers and tenants while working to save money to purchase their own land.
From *The Soldier in Our Civil War*

END NOTES

Matt Gardner Homestead

West, Carroll Van. "Matt Gardner Homestead, Giles County, Tennessee." National Register of Historic Places Nomination Form, Tennessee Historical Commission. Listed 1995.

McDonald Craig Farm

Woodcock, Jaime. "Craig Family Farm." National Register of Historic Places Nomination Form, Tennessee Historical Commission. Listed 2005.

Robertson Farm

Robertson, Evelyn C., Jr. *Education and the American Dream: The Allen-White High School Story 1905-1970* (Charleston, South Carolina, 2009).

Szcodronski, Cheri LaFlamme. *Finding Freedom at Grand Junction, Tennessee: Ulysses S. Grant, Chaplain John Eaton, Jr., and the Contrabands.* Masters Thesis (Murfreesboro, Middle Tennessee State University, 2011).

Refer to the "Bibliographical Essay" for additional resource materials.

Rebuilding the Farm

When the Civil War ended in the spring of 1865, those who fought and those who survived turned to yet another daunting task -- rebuilding lives and farms within a devastated land. The veterans found farms in ruins with animals missing and crops not planted. People were hungry, and prospects were bleak for former soldiers and their families. A return to successful or even subsistence farming was often impeded by physical infirmity, the scarcity of farm labor, little money, and the lack of livestock, supplies, and tools. The generals and armies were gone, but the bitterness and hatred of 1861-1865 lingered in many places. Reprisals and violence remained far too common; some disputes escalated into feuds that scarred communities and families for decades.

Historian Robert Tracy McKenzie has studied the postwar reorganization of Tennessee agriculture and reports that an 1872 Congressional committee estimated the state's wartime losses at over $185 million. This figure included both slave and non-slave property, with at least $96.5 million attributed directly to the emancipation of the state's 275,000 slaves. The non-slave property losses were approximately $89 million, roughly one-third of the total value of property recorded in 1860. Rebuilding houses, farms, families, and communities, as well as roads, railroads, institutions, commerce, and industry, was indeed a "long row to hoe."

The term "Reconstruction" generally applies to the years between the end of military hostilities in 1865 and 1877 when President Rutherford B. Hayes removed Federal troops from the last of the southern state capitals. Tennessee rejoined the Union in 1866 and escaped most of the military program applied by Congress to the other former Confederate states. Times remained troubled, however, for most Tennessee residents as they coped with the realities of change. In the many families where male members did not survive the war or returned too wounded to work, women and children continued to carry the burden of farm management and daily operations. Some former slaves, as well as whites who were landless, became tenants or sharecroppers. For those who could afford land, establishing a new farm meant a new beginning. One in four farms that are certified Century Farms today were established from 1865-1880. In addition to farms established by veterans and ex-slaves, other were founded by immigrants from western Europe who launched their own hopes on Tennessee's post-war landscape. These new settlers brought their traditions and methods of farming with them, along with a variety of barn and house styles.

Family farms after the Civil War, in some ways, operated much as they had prior to the war, using the same methods and tools to raise livestock and crops. But in many other ways, the last 35 years of the nineteenth century introduced new chapters into Tennessee agriculture. Many families turned to the introduction of new breeds of livestock, different row crops, and improved technology. Most importantly, slavery was no more, but many land owners were still dependent on black laborers, now tenants, sharecroppers, or hired daily wage-earners, to do most of the hard work. The emergence of agricultural boards and fairs promoted and assisted farmers across the state as survivors of the Civil War focused on rebuilding their farms and their communities.

Cleburne Jersey Farm
Maury County

The Cleburne Jersey Farm in Spring Hill, near the Maury and Williamson county lines, illustrates in its buildings and history the significant change in direction that family farmers took following the Civil War. Originally part of the Campbell plantation, one of Middle Tennessee's largest, the Cleburne Farm was established in 1872 by McCoy "Mack" Campbell. He named it in memory of Confederate Gen. Patrick R. Cleburne, who fell at the Battle of Franklin in 1864.

The 1870s house and original dairy, just to the rear, of the Cleburne Jersey Farm are listed in the National Register of Historic Places.

Campbell was the nephew of Lizinka Campbell Brown Ewell whose father, George Washington Campbell, had served as a U.S. Senator and minister to Russia. Lizinka had inherited the Maury County property from her father and brother. A wealthy widow and prominent in Nashville society, Lizinka was actively involved in the management of the farm, though overseers handled its daily operations and enslaved people did the work.

When the Civil War began, Lizinka's world changed completely. As a strong supporter of the Confederacy, she organized and led the Ladies Hospital Association in Nashville and, in 1861, helped outfit the "Brown Guards" of the 1st Tennessee Infantry Regiment organized by her cousin George W. Campbell, who served as its captain. Mack Campbell joined as an officer. Lizinka's son, Campbell Brown, who was about the same age as his cousin Mack, became a lieutenant in the 3rd Tennessee Infantry. When Federal troops occupied Nashville in 1862, Lizinka Campbell Brown left her Nashville mansion for Richmond, Virginia. The Campbell-Brown mansion became the official residence of Military Governor Andrew Johnson.

Lizinka Brown married her first cousin, Gen. Richard Henry Ewell, in May 1863. Ewell was raised in Virginia, graduated from the United States Military Academy in 1840, and served under Zachary Taylor in the Mexican War. During the Civil War, he fought in several major battles, including Gettysburg, was wounded on at least two occasions, and was imprisoned towards the end of the war. Lizinka Ewell gained amnesty for her husband and son by appealing directly to President Andrew Johnson, who agreed to the pardons in 1865. Mrs. Ewell also worked diligently to regain possession of her farm which had been classified as an "abandoned plantation" in her absence. Though it took some time, she was successful in her efforts. By 1867, the couple occupied the property and renamed it "Ewell Farm." Cousins Brown and Campbell soon joined them and became the active managers under the direction of the Ewells.

For the next five years, the foursome began to plan and build an agricultural legacy. Ewell, Brown, and Campbell persuaded Lizinka to purchase Jersey bulls from a Maryland herd. With these two bulls, they began to experiment and improve their livestock, although they did not begin dairying on a large scale for several years. With the deaths of the elder Ewells in 1872, the farm passed to Campbell Brown, who immediately sold 310 acres to Mack Campbell. Each of the families built a substantial house on their property.

The origin of the Jersey dairy industry in Tennessee dates to 1878 when Campbell Brown, with assistance from Mack Campbell and William J. Webster, purchased purebred cows to mate with the bulls they had bought a few years earlier. By the 1880s, the Cleburne Jersey Farm's herd had greatly increased, and it supported a profitable dairy operation, with surplus Jersey stock for both breeding and sale.

Mack Campbell relied on African American tenant farmers and hired hands to do the hard work of twice-daily milking, breeding livestock, and growing crops like tobacco. One tenant who lived on the farm for many years was Will Barnes. He grew up on the property as a slave and became a respected and a knowledgeable farmhand.

The Cleburne Jersey Farm attained a national and international reputation as a progressive and profitable agricultural enterprise. In 1927, the editorial staff of the *Jersey Bulletin and Dairy World* magazine recognized the Cleburne Farm as the oldest continuously operating Jersey herd in the United States. The Campbell family continued to improve and enhance the reputation of Cleburne Farm as a pre-eminent stock farm through the first half of the twentieth century. Today, the Campbell family continues the long-standing tradition of raising and milking Jerseys that began with the combined vision of three Civil War veterans, an enterprising woman of means, and the labor of African American tenants who had once been slaves. This story of agricultural transformation and transition was repeated in similar ways across the state.

The Cleburne Farm has the oldest continuously operating Jersey herd in the United States.

Maplewood Farm
Williamson County

Within a few miles of the Cleburne Jersey Farm, the Lee family of Maplewood Farm in southern Williamson County also relied on livestock, as well as diverse crops, to help re-establish their agricultural operation after the Civil War. Samuel Brown Lee, of Connecticut, was an adventurous young man, and the only one of his siblings who elected to travel to the frontier in the 1820s to the 5000 acres deeded to his mother, Elizabeth Brown Lee, by her brother, Daniel Brown of South Carolina. Only eighteen, Samuel Lee made the long journey south and across the mountains to claim the land and build a log cabin. Early in his Tennessee residency, Lee became involved in the emerging iron business, living and working in Memphis for a time before returning to his farm. By the early 1830s, Samuel held a substantial interest in the Duck River Furnace and, in 1837, he married Susan Amanda Napier, whose family was one of the most prominent in the antebellum iron industry. After the Duck River enterprise became less economically viable, and he was often away from home on business, Lee wrote to his wife that "I should just be a farmer." He did become a leading planter and was particularly known for his high quality mules. In 1850, his farm was valued at $15,000, with $2600 of that total allotted to livestock including thirty five asses and mules. Lee valued his mules highly and drove them himself, with the help of a slave, to sales in Mississippi.

By 1860, Lee owned twenty six slaves, and his farm was valued at $35,000. In that year he began growing tobacco rather than cotton as a principal cash crop. He also increased Indian corn production to 4000 bushels and wheat to 1300 bushels. When the Civil War began, his three sons joined the Confederate army and rode with the 4th Tennessee Cavalry. The farm's location, near Spring Hill, and its productivity made it vulnerable. Foragers from both armies took stores and livestock, burned fences, and the house was used as a hospital. Many of the slaves left to seek freedom behind Federal lines. By the war's end, though most of the buildings had survived, the landscape and the farming operation were in ruins. Samuel Brown Lee died in June 1865, leaving his sons to rebuild the farm.

Samuel Brown Lee, Jr., John Wills Napier Lee, and Charles Lee returned from the war and, with the help of tenants, began to revive the farm. Family history relates that, at first, the only livestock they had was one lame mule and the worn mounts from their cavalry days. The Lees operated at a much smaller scale than before 1860 and changed their farming strategy. By 1870, they grew no tobacco but began producing hay, corn, and wheat, and some small amounts of cotton again. Poultry was also important, with John collecting about two hundred dozen eggs in that year. Charles had five horses and seven mules, while John had thirteen mules and six horses. About

1872, however, John Lee acquired a horse which helped to increase the reputation and the fortunes of Maplewood. "Duplex" won the World's Championship Pacing Record and became a nationally known sire of thoroughbreds.

As the nineteenth century progressed, Maplewood was known for the breeding of thoroughbred horses, though a few mules were kept for the diverse farm work. Samuel Lee, Jr., who operated about 40 acres in 1880, served in the state senate from 1897-99. The long history of the Lee family of Maplewood Farm, who continue to live in the 1830s house has been well-documented over the years, and the farm is listed in the National Register of Historic Places for its architectural and agricultural significance.

(Opposite page) Civil War veteran John Wills Napier Lee is seated with his granddaughter, Eunice, standing next to him. On the back row are his children Bess, John Wills Napier Lee, and Eunice Mallory. The circa 1835 house in the background is home to current farm owner John N. Lee and his family.

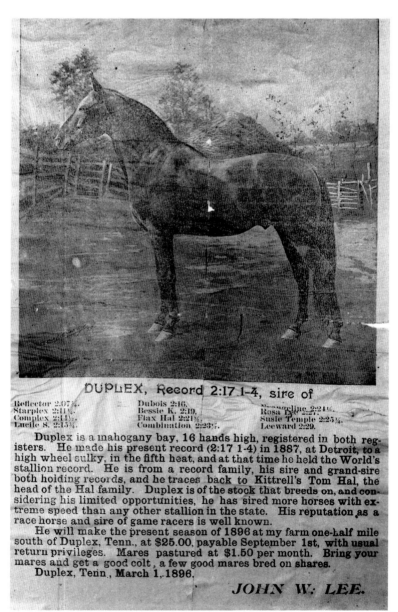

This rare handbill from 1896 may be the only image of the famous Duplex.

Lancaster Farm
Smith County

Located on the banks of the Caney Fork River, the Lancaster Farm, with more than 2500 acres, had contributed to Tennessee agriculture for more than seventy years when the Civil War erupted in 1861. Founded by John Lancaster, Jr., in 1790, it predates Tennessee's statehood by six years. As the war began, the fourth-generation owner, William Lancaster, was in the process of building a handsome two-story home for his wife, Elizabeth, and their three children. Construction had to stop, however, and the house was not completed until after 1865. Family tradition holds that William freed the 11 slaves he owned in 1860 and provided them a place to live.

In 1867, Melissa, one of two Lancaster daughters, married Confederate veteran James C. Prichard, also a native of Smith County. Prichard enlisted in Company F, 24th Tennessee Infantry (CSA) in July 1861. He was wounded at the Battle of Stones River, captured at Missionary Ridge, and imprisoned at Rock Island and later Richmond, Virginia. After his release, he went to South Carolina and was with Gen. Johnston's army at the time of surrender in 1865. James and Melissa Prichard acquired 309 acres of her father's farm in the fall of 1867. The Prichards eventually owned other portions of John Lancaster's original holdings on which they raised wheat, corn, tobacco, swine, and poultry.

Railroad development changed the farm's history in the 1880s when the Prichards sold 400 acres to the Nashville & Knoxville Railroad Company. The railroad built a flagstop at Seabo Wisha on the property, which also included buildings and a club house for use by railroad officials.

Smith Fork Creek, which forms a boundary for part of the Lancaster Farm, is a tributary of the Caney Fork River, one of Tennessee's most important and scenic waterways.

These buildings were located near the existing railroad bridge crossing the Caney Fork River. The couple successfully shepherded the family farm through the years following the war and into the twentieth century. Melissa Lancaster Prichard was 88 when she died in 1921, and James Prichard died at the age of 96 in 1936; both are buried in the Prichard Cemetery.

J. M Bailiff Farm
DeKalb County

Typical of the many young men and women who chose to return to their pre-war rural lifestyle were the Bailiffs of DeKalb County. James Monroe Bailiff and his brother, Columbus, supported the Confederacy and joined Col. R.D. Allison's Cavalry Squadron in February 1863. Allison's three companies actively patrolled DeKalb County and took part in the Battle of Snow's Hill on April 2, 1863, which occurred about a mile from the farm that Monroe Bailiff would purchase after the war.

Allison's squadron fought in a number of battles and was attached to different commands, including those of Generals Joseph Wheeler, Nathan Bedford Forrest, and Frank Armstrong, as well as Lt. Gen. Wade Hampton's Cavalry Corps at the end of the war. It was, however, during the Battle of Chickamauga in Georgia in September 1863 that Monroe Bailiff was wounded and Columbus Bailiff fell ill with typhoid fever. A local family took them in and nursed them as their unit retreated southward. Soon Federal troops captured both wounded men and sent them to prison in Louisville, Kentucky. In ill health and unable to perform additional military service, the brothers were released in the spring of 1864 and made their way back to DeKalb County. Having survived imprisonment and his wounds, the nineteen-year-old Monroe Bailiff wanted nothing more than to make a living and raise a family. He married Eliza Foster from the nearby community of Wolf Creek in October 1865. It would be another decade, however, before the couple could purchase a 52-acre wooded hillside property in Possum Hollow in the Dry Creek community.

With three small children, the Bailiffs completed a partially built log house and cleared the land of trees with a team of oxen. During the next ten years, following their land purchase in 1875, the oxen were sold and replaced with mules, which were better suited to hillside farming. Bailiff and his sons worked long hours to produce corn and wheat crops. They also raised poultry, bees, hogs, and cattle. Bailiff was an experienced third-generation blacksmith. He was also skilled in leather work and repaired shoes for members of the community. The family, which eventually included nine children, made steady progress to secure the land and a future for the succeeding generations who would live on the farm.

James Monroe and Eliza Bailiff worked hard to save money to purchase their farm in 1875.

Woodard Hall
Robertson County

Some farm families returned to the products that had served them well in antebellum days but they necessarily made some changes to accommodate the postwar economy. Woodard Hall in Robertson County, dating to 1792, is one of the oldest farms in the state. Before the war, Wiley Woodard, born in 1810 and the second-generation owner, increased his holdings to 2,000 acres on which he cultivated row crops with the help of slaves; according to the 1860 slave census, twenty six enslaved people lived on the plantation. With the additional help of an Irish distiller, Wiley Woodard also directed the expansion of the successful distilling operation begun by his father, Thomas. The mash from the distillery fed a large swine herd that filled the smokehouse with ham, bacon, and sausage.

Woodard realized that improved transportation was a key to his plantation's success and, as a member of the Tennessee legislature from 1849 to 1853, he sponsored legislation for both the Manscoe's Creek & Springfield Turnpike, and the Edgefield & Kentucky Railroad, both of which passed within three miles of his property and were completed before the Civil War. The distillery's output increased as Woodard marketed whiskey and apple and peach brandy to customers in Louisiana, Mississippi, Alabama, Kentucky, and New York.

Because of this commercial success, Woodard was able to expand the farm house into the substantial dwelling that exists today. Other buildings dating to this prosperous period before the war include a ca.1854 brick kitchen, the original log smokehouse, and the farm office. In 1860, farm sales reached $14,850. Though the Civil War interrupted production at the distillery, it resumed quickly once the armies left. By September 1865, Woodard had shipped 249 gallons of whiskey to Nashville, and within ten years, Wiley Woodard & Company was earning nearly $75,000 in annual revenue.

By 1870, Woodard's land was valued at $21,000. Among his animals were ten horses, six mules, and a swine herd of fifty head. His livestock was valued at over $3,000.00. Woodard moved to the post-bellum practice of using tenant farmers to take care of the livestock and plant and harvest row crops, which included 1,200 bushels of Indian corn and 100 bushels of sweet potatoes. Some of the workers had been slaves at Woodard Hall prior to emancipation. When Woodard died in 1877, he bequeathed land to three of his former slaves, Charry, Elizabeth, and Amanda, and set aside money to build houses for them and their families. These houses were constructed in the same area where the slave quarters once stood. The cemetery on the property contains the graves of generations of the Woodard family within a walled enclosure. Beyond it is the burying ground of slaves and tenants who lived at this plantation. Woodard Hall is listed in the National Register of Historic Places.

Woodard Hall's appearance today is much the same as it was in the 1850s.

(Opposite page) Broadside of Wiley Woodard

Col. Wiley Woodard
(1810 - 1877)

WILLIE WOODARD,

DISTILLER OF

Pure Robertson County Whisky,

APPLE AND PEACH BRANDIES,

Springfield, Robertson County, Tenn.,

Terms Cash.   Orders Solicited.

Samples Sent Free.

REMEMBERING VETERANS

In the years following the end of the Civil War, former Union and Confederate veterans and sympathizers began to create organizations to recognize soldiers. These groups sponsored reunions and other opportunities for social gatherings, and provided formal commemoration which was their most important purpose. The United Daughters of the Confederacy erected monuments and awarded medals to veterans. Likewise, the Grand Army of the Republic, awarded valor, but saw their purpose change over the decades as the group became less political and more focused on reunions and pension reform.

Grand Army of the Republic

The Grand Army of the Republic (GAR) was a Union veterans organization first formed in Illinois in 1866. Local groups of the organization were known as "camps" and members of the first camps in Tennessee were largely African-American veterans and federal employees that supported Radical Reconstruction. Reaction to the GAR was hostile in some parts of the state during the 1860s. An article in the Clarksville Weekly Chronicle from April 12, 1867, described the GAR as "manipulated by those yankee adventurers who, with or without ostensible business, infest every portion of our territory." The article went on to state that white members of the group would induce African Americans "to swear eternal hostility to the white men of the South." The Memphis Daily Appeal wrote on July 5, 1869, that the state had "been robbed, plundered and well nigh bankrupted" by the GAR.

No doubt, due to the hostility and suspicion described above, many of the GAR camps in the state were defunct by the end of the 1870s. The 1880s, however, saw a revival of the GAR as the group increasingly drew white males and embraced Southern racial attitudes. GAR camps held reunions with Confederate veterans and often marched with former Confederates in parades. By 1891, there were more than 3,700 members of the GAR in Tennessee, and a total of 400,000 in the organization. Camps and posts were found across the state in towns including Nashville, Rover, Greeneville, Pulaski, Loudon, Clarksville, and Pikeville. By 1890, Memphis had a chapter of the GAR that met on the first and third Tuesday of the month. Chattanooga hosted the 47th annual National Encampment of the GAR in May 1913, showing that the earlier hostility against the group had all but disappeared.

The Grand Army of the Republic monument in Cleveland's Ft. Hill Cemetery is one of three similar monuments in Tennessee.

In the course of its nearly 100 years history, the GAR founded soldiers' homes, and was active in pension legislation. Presidents Grant, Hayes, Garfield, William Henry Harrison, and McKinley were members of the GAR. As Union veterans began to die, the GAR's membership declined each year, though such groups as the Sons of Union Confederate Veterans of the Civil War assumed some of programs. The GAR existed until the death of its last member, Albert Woolson, in 1956. At that time, the GAR officially dissolved.

Markers of Grand Army of the Republic and the Southern Cross are displayed at the grave of an unknown Civil War soldier in Rest Haven Cemetery in Franklin

Southern Cross

In the late 1890s, the United Daughters of the Confederacy began a program to honor veterans. The idea is attributed to Mrs. Alexander E. Erwin who was inspired while attending a reunion of Confederate veterans in 1898. The cross is bronze with the motto "Deo Vindice", and the seal of the Confederate states on one side. The reverse side holds the battle flag with thirteen stars. The rules governing the bestowal of the cross were strict, and required positive proof of honorable service to the Confederacy.

Charles Clayton Abernathy, Jr., of Crack Hill Farm in Giles County, also received the Southern Cross of Honor. He served as a commissioned surgeon in the Confederate army from 1862-1865. The diary kept by his wife, Martha Abernathy, provides the story of Dr. Abernathy's war years. He was a prisoner of war three times during the war. He first left Pulaski in June 1862, as a prisoner after refusing to sign the Oath of Allegiance to the United States. He was later captured near Jonesboro, Georgia, on September 1, 1864 and was part of a prisoner exchange on September 28, 1864. His final capture was in Pulaski on December 25, 1864. He spent time in multiple prison camps prior to his release on July 19, 1865. Dr. Abernathy also served as the president of the Tennessee State Medical Society, the Giles County Medical Society and the Pulaski Medical Society.

Richard A. Mann of Haywood County received the Southern Cross years after the end of the Civil War. At age 16, while still in high school, he left home, riding his sorrel horse, Gem, to join the Confederate army. He was accompanied by his brother who was killed during the war. In the 31st Infantry, Company D, Mann fought at Shiloh and other battles and skirmishes. Ten years after Appomattox, Richard Mann, by then married to Harriet Taylor, purchased a farm north of Brownsville. The owners of the Johnson Farm remain stewards of the land and the medal awarded to their ancestor.

Richard A. Mann, recipient of the Southern Cross, and his wife, Harriet Taylor Mann, were among many couples who established a farm in the decade after the Civil War.

Brabson Ferry Plantation
Sevier County

John Brabson II came from Virginia in 1794 and founded a farm near the French Broad River at Boyd's Creek in the area that would become Sevier County, North Carolina, in that same year. In 1826, he was granted permission by the Tennessee General Assembly to construct a dam on the south side of the river near the head of Boggs Island. Brabson, married to Elizabeth Davis Brabson and the father of ten children, was an enterprising farmer and merchant. He acquired more acreage, and he operated mercantile stores, blacksmith shops, tanneries, and carpentry shops. The farm produced a variety of grains and livestock. When John Brabson died in 1848, his will disclosed forty six slaves, one of whom he freed at his death.

Benjamin Davis Brabson became the owner of much of the plantation in 1848. He and his wife, Elizabeth Berry Toole, had eight children. Like his father, Benjamin was as entrepreneur and astute businessman. With his brother he established Brabson and Brother in 1852, and they ran a complex of businesses along with the farming operations until the Civil War. Benjamin built a house in 1856 that remains the primary residence of the family today.

William Toole Brabson, a son of Benjamin and Elizabeth, rode with L Co, 8th Tennessee Cavalry (CSA) and was captured at the Battle of Chickamauga. He was imprisoned, along with his cousin, John E. Davis, at Camp Morton, Indiana. Benjamin Brabson wrote to President Lincoln to request the return of his son and nephew. Apparently,

Glen Villa was built in 1856 by Benjamin Brabson.

no response was received as William remained in prison until the end of the war. By that time conditions in Sevier County were described by the family as "intolerable" for Confederate sympathizers, and many people moved to safer places, at least for a time. Benjamin Brabson moved to Winchester, Tennessee, and died there less than a year after the move. His mother, then 74 years old, wrote him a letter on October 14, 1865, describing the family's situation after he left. She told of her own journey to Knoxville from Sevier County after she was "run from her home" on the night of September 22. She reported that "they rocked my house so, burned my carriage, and fired my outbuildings, they also shot at me several times." After the situation was reported to Federal authorities in Knoxville by a family member, arrangements were made for Elizabeth Brabson to be brought to safety. She was provided an ambulance, three wagons for her belongings, and a guard of "six colored soldiers" to escort her to Knoxville. Elizabeth Brabson died in Knoxville in 1868, but her remains were returned to Sevier County and buried in the Brabson Cemetery.

After Benjamin Brabson's death, family members, including William Toole Brabson, began returning to their land and homes. By adopting stringent economic measures and with the help of former slaves, some of whom remained as tenant farmers, the Brabsons began to cultivate land, raise livestock, and resume a diverse operation. Although most of the family's household goods were lost in a warehouse fire in Knoxville in 1865, the farm buildings remained intact, so the family was in a better position than some of their neighbors. In addition to the farm house, early nineteenth-century log outbuildings remain as does a circa 1875 tenant house and an 1880s barn. Brabson Ferry Plantation is listed in the National Register of Historic Places.

Along the farm's bottom land is the site of the ferry crossing that gave the plantation its name.

Spring Meadow Farm
Knox County

In the Strawberry Plains community of Knox County, the Reverend David Adams, a circuit-riding Methodist minister, established a farm in 1842. After his death in 1853, a portion of the property was deeded to his daughter, Amanda, who married Joshua Curtis Bailey in 1856. The Adams and Bailey families became intimately acquainted with the hardships of war. Joshua's brother, Ike, was in the Confederate army, as were Amanda's two brothers, Samuel and Thomas. Samuel Adams died from wounds received in the Battle of Seven Pines near Richmond, Virginia, in early 1862.

The Bailey farm was a short distance from the 1,600-foot Tennessee & Virginia Railroad bridge that spanned the Holston River. Union and Confederate forces fought for control of the bridge during the winter of 1863-1864. Strawberry Plains Academy, of which Rev. Adams had been a founding trustee, was just a mile from the farm and was destroyed during the fighting.

After the war, the Baileys studied developing trends in agriculture to take advantage of recently introduced livestock breeds and new varieties of row crops. The progressive farmers began to rebuild and added another 66 acres in 1873. Joshua introduced orchard grasses for pasture and hay. He also sold seed to farmers as far away as Arkansas. Joshua Bailey was a pioneer in soil conservation by planting winter crops and cover to prevent soil erosion. Spring Meadow Farm is an apt name for this historic property that remains in the Bailey family.

The Bailey family knew first-hand about the fighting that took place in the winter of 1863-1864 to control the Tennessee & Virginia Railroad bridge at Strawberry Plains near their farm.

Courtesy *Library of Congress*

Joshua Curtis Bailey was a pioneer in new agricultural methods and crops after the war.

POST-WAR COUNTY FORMATION

The fifteen years following the Civil War produced a reordering of boundaries in Tennessee, and the state's leaders created twelve new counties between 1870 and 1879. The initial burst of activity occurred with the 1870 Constitutional Convention. One of the purposes of the convention was restore power to disenfranchised elements of white society, and delegates worked to remove the state from Unionist control and restore former Confederates to as many positions of power as possible. The 1870 constitution also called for the formation of Clay County, Hamblen County, Lake County, Loudon County, and Trousdale County. A year later the Tennessee General Assembly authorized Crockett County, Houston County, Moore County, and James County. Unicoi County was established in 1875, and Pickett and Chester counties were carved from existing boundaries in 1879.

Geography was often cited as the reason for creating new counties. Residents in what is now Houston County petitioned for their own county because they felt they were too far from the courthouses in Dickson, Humphreys and Stewart counties to conduct business or take part in county affairs. Residents in Lake County cited the same geographic concerns, noting the difficulties in crossing the "scatters of the Lake," a swampy area south of Reelfoot Lake to attend court in Obion County. Richard Stanford Bradford was one of the organizers of the new Lake County in 1870. He had established a farm of 140 acres in 1861 and was an official in Obion County prior to the new county's formation. Bradford was the first magistrate of the Third District and served as chairman of the Lake County court for many years. His farm grew to 2,700 acres, and was later named the Joe Carter Farm after one of his descendants.

In their petition for county formation, residents in Moore County also pointed out the distances to the county seats of Bedford, Coffee, Lincoln and Franklin counties to conduct business, as well as the terrible road conditions. Those in what is now Clay County argued that they would have more access to self-government if they had their own county, noting that there were few roads and trails in Overton and Jackson counties. Hamblen Countians also noted the difficulties of reaching the county seat due to the lack of roads and the terrain of Jefferson and Grainger counties from which their county was partly formed in 1870.

Politics, of course, also played a role in the formation of new counties. Residents of what is now Crockett County first voiced a desire for more convenient access to county government in the 1830s, and the county unsuccessfully petitioned the state for its formation in 1832. The state did not authorize the county until 1845; a year later, the state reversed the decision, stating that Crockett County, named for frontiersman David Crockett, was not a constitutional county. It was not until 1871 that the state legislature authorized the creation of Crockett County. A similar story is that of Wisdom County, officially designated as such in West Tennessee in 1875. Never organized, the county languished until 1879 when the state reorganized the county and called it Chester County.

James County was organized in 1871 from Hamilton and Bradley counties. Political machinations were a part of the creation of the county; citizens of the James County area were largely Republican and rural and did not wish to remain in the same county with the largely Democratic and urban residents of Chattanooga (Hamilton County). From the start, troubles plagued the county; it was thinly settled and political strife continued through its brief history. In 1919, the county went bankrupt and was dissolved. Since that time, Tennessee has operated with 95 counties.

Counties Created in the 1870s

County	Year	County	Year
Clay	1870	Lake	1870
Chester	1879	Loudon	1870
Crockett	1871	Moore	1871
Hamblen	1870	Pickett	1879
Houston	1871	Trousdale	1870
James	1871	Unicoi	1875

Whitetown Acres Farm
Hamblen County

Holding slaves did not necessarily mean that farm families were in complete sympathy with the Confederacy. Unionist Jimeson White oversaw a diverse farming operation, with the help of slaves, on land that had been in his family since 1812. A letter written by White in the mid-1870s to his cousin, Jonah White, provides some insight into the family's wartime experiences. "The war was very hard on us as we was on the playground between the contending partys and community was about equal divided on the war question," White recalled. The farm was within three miles of the East Tennessee & Georgia Railroad, so both armies foraged regularly in the area. White went on to say that he was a "Union man through the war," but his brothers Joseph and John "was Suthern men though not streanious." Along with news about various family members, he disclosed that two sons had fought for the Confederacy while his third son, Daniel, died in the Federal service in Nashville in 1863.

Though he lost his buildings, possessions, and fences to troops and foragers, White was able to keep 200 acres. After emancipation, White remarked that some of the freedmen remained nearby and worked in the house and on the farm. Within a few years, and with their help, White had rebuilt outbuildings and fences and had "plenty of this worlds goods." He reported the prices of livestock and crops of corn, wheat, and oats. White had a good market for his produce at the railroad. His son, William, operated a saw mill, a corn mill, a distillery, and with his brother, Washington, had invested $600 in a threshing machine with which they were "making money." Jimeson White went on to say that the "county is generaly helthy as far as I know" and closed his postwar report by asking his cousin to write and give him a report on his family and how they "made it through the war."

Jimeson White, though a slaveholder, was loyal to the Union.

Shady Oaks Farm
Hamblen County

Providing educational opportunities for former slaves concerned the Reverend William C. Graves, even though he and his family had sided with the Confederacy. This conflict between philosophy and reality was one several individuals and families struggled with before, during, and after the Civil War. A minister of the Methodist Episcopal Church (MEC) beginning in 1836, Graves joined with other pro-slavery dissenters who parted with the denomination in 1844-1845 to form the Methodist Episcopal Church, South. From 1858 until 1861, Graves was the editor of the *Religious Intelligencer,* a weekly newspaper published in Morristown by Neal & Barnett. In 1860, Graves purchased just over 100 acres near present-day Morristown. He and his wife, Martha Washam Cardwell, were the parents of ten children. During the Civil War, two sons, Jason Perrin and Thomas Alexander, enlisted in the 61st Tennessee Mounted Infantry Regiment (CSA). Family tradition holds that the Graves family offered shelter to both Union and Confederate soldiers.

When the fighting ceased, Graves was "desirous of peace." In the *Minutes of the Holston Conference of the Methodist Episcopal Church, 1896,* Graves is described by J. S. Jones, also a Methodist minister, as a "peacemaker; bending all his energies to promote harmony between his Methodist brethren" who were divided by the war. When the Methodist denomination reorganized after the war, Graves was among the first to "re-enter her fold and labor for her success."

Jones wrote that the best work of W. C. Graves was his longtime support of the Morristown Normal Seminary. Founded by the Freedman's Aid Society of the MEC in 1881, the school educated "colored men and women as teachers and preachers for the colored work in the South." When Graves died in January 1896, Jones, who conducted the funeral services, reported that the "students of that school showed their appreciation of their old friend by walking four miles to attend his funeral in a body." Graves was survived by five of his ten children, including William H. B. Graves, who acquired the farm in 1872, just two years after the formation of Hamblen County. Charles E. Graves is the current owner and his son, Charles David Graves, works the acreage.

The Reverend William C. Graves helped to raise funds for Morristown Normal Seminary, a school for blacks, after the Civil War.

In **Fentress County,** on the upper Cumberland Plateau with Kentucky as its northern boundary, residents were largely pro–Union. Jacob Wright and his wife, America, were the second generation to farm acreage purchased by his father in 1835. Their sons were too young to serve in the army, but the family "hid valuables and food from both Union and Confederate troops" who were foraging in the area. The family dates this corn crib and barn, which "still has original pegs and seals," to the period just after the Civil War when Wright began to rebuild his property. At the farm, now called the **Cooper Farm,** the eighth-generation descendants of Jacob and American Wright continue to use these outbuildings in their farming operation.

Cypress Creek Farm
Benton County

Cypress Creek flows through the farm by the same name on its way to the Tennessee River. William and Penelope Thompson settled here soon after the Chicksaw Indians ceded all of West Tennessee to the United States in 1818. A stretch of the main stage route, referred to at various times as the Camden Road, Camden-Rockport Road, or Camden-Decaturville Road, was called Thompson Lane as it passed through their land.

When William died suddenly in 1842, he left a tract to each of his 12 children. His son and the executor of his estate, Charles Hodges Thompson, eventually purchased several tracts from his siblings to make a sizeable farm. On it he built his own log house in the 1850s, not far from his mother's home, where he and his wife, Fredonia "Adaline" Utley, and their ten children lived. Engaged in a wide range of agricultural activities, the Thompsons planted an orchard, kept bees, raised livestock, and cultivated corn, wheat, tobacco, and vegetables. In 1860, Thompson built a bridge on the Camden-Rockport Road for travelers to cross Cypress Creek for which he received $20 from the county court. In that same year, the farm produced 250 bushels of Indian corn along with a few bushels of wheat. The family owned two horses and several head of cattle and sheep as well as swine. Some of the land was tenanted to Charles Thompson's brothers, Jerry and Dave.

This log house was built by farm founders William and Penelope Thompson in 1819. Charles Hodges Thompson was the first of their children to be born here. The house was listed in the National Register of Historic Places in 1976.

During the Civil War, the Thompsons remained loyal to the Union, but "had to be on constant alert for danger because there was so much indiscriminate killing from Confederate and Unionist elements, plus the outlaw gangs," the family recalls. Charles was a target for local secessionists who, on one occasion, set fire to the corn crib. When he rushed to put out the flames, he "would have been shot if not for one in the crowd that prevented it," writes descendant Carol Branch in her history of the farm. Adaline was forced to "slice and cook a middlin of meat for the raiders." At the time of Gen. Nathan Bedford Forrest's raid on Johnsonville (just across the Tennessee River) in 1864, troops raided the farm of Dave Thompson, brother to Charles, and took 200 bushels of corn and 500 bushels of fodder. After the war, Dave Thompson filed a claim for damages and, with his brother Charles and neighbor Edward Branch as witnesses, his application was approved and he received $212.50 in compensation.

Although the family had a productive farm following the war, taxes and debt hit the Thompsons hard when Charles died suddenly in 1879. The land was forced into foreclosure and auctioned. Adaline Thompson's uncle, Henderson Taylor, purchased much of the property and immediately deeded 80 acres back to her. She and her daughters, Nancy "Velonia" and Arena Jane, likely with the help of other members of the family, plowed, cultivated, and harvested their crops for market and for their table and livestock. Adaline Thompson died in 1896, just one year after her mother, Penelope, died at age 97.

Fredonia Adaline Utley Thompson and her daughters are credited with securing the family farm after regaining it with the help of her uncle, Henderson Taylor.

Fredonia Adaline Utley Thompson

Walnut Flat Farm
Madison County

In 1834, William and Eliza Pearson with their two young sons moved to Henderson County near the Parker's Crossroads community. By 1860, Pearson had increased his landholdings to a sizeable plantation of 2,000 acres, and he owned forty slaves. One of the region's major planters, he directed his slaves in cultivating cotton, corn, oats, and wheat and in raising a large herd of cattle and other livestock. Pearson's real estate was valued at $10,500, and his personal wealth was recorded at $52,700 in the year before the war.

Parker's Crossroads was the scene of a major battle late in 1862. Surviving the battle, however, was not the only challenge for the Pearsons. Federal soldiers once stopped at the Pearson house, "tied their horses to the columns, and with the help of the slaves freely looted and took whatever they pleased." With their home and property in shambles, and grieving over the death of their son, "Little Billy," in the Battle of Atlanta in 1864, the Pearsons moved to a farm in Madison County in 1865 and began to rebuild their lives. Pearson eventually expanded his property to 1,357 acres, and he returned to growing cotton, corn, and hay, as well as raising livestock. Pearson and his surviving sons purchased over 4,000 acres of the fertile Butler Plantation land that was divided into tracts and sold after the war. They promised to pay $4,200 by installments for this property. The story has long been told in the family that the "sharpies" who sold this property at an inflated postwar cost felt sure that the family would default on the payments. The Pearsons, however, made all the payments on time.

The Pearson family also helped to rebuild their new community. In 1870, William and two sons, Jonathan Dudley and John Solomon Pearson, donated land for the construction of the Claybrook Cumberland Presbyterian Church. On March 17, 1875, the Pearsons, father and sons, deeded a fourth of an acre, east of Claybrook Road, to Barns Meachum, Sampson Willis, Robert Cox, and Charles Anderson as trustees for the Colored Methodist Episcopal Church for a church site. Then in September 18, 1885, the Pearson brothers made a clearer title for slightly more land for this church.

A daughter of William and Eliza Pearson, Susan Elizabeth, inherited 153 acres of the plantation in 1871. Susan married John George Woolfolk ,and this couple's lineage and name continues in the current owners. The historic farmhouse dates to 1870, and is a symbol of the family's new prospects following the hard years of war.

John George Woolfolk and his wife, Susan Pearson, farmed 153 acres of her family's plantation that she inherited in 1870.

Estanaula Oaks, Hilltop, And Harvey Hill
Fayette County

In 1853, Dr. Richard Henry Harvey purchased just over 200 acres eight miles north of Somerville, the Fayette county seat, on the Somerville-Stanton Road. Harvey practiced medicine but was a successful planter in a county that had some of the largest plantations in the state. Over time, he increased his acreage from which three Century Farms evolved—Estanaula Oaks, Hilltop, and Harvey Hill. Harvey supervised slave labor to produce a large cotton crop as well as corn and livestock. In 1860, Harvey owned twenty two slaves including males, females, and children. Harvey and his first wife, Dillia Ann Hammons, were the parents of seven children. With his second wife, Mary Jane Rogers, he had one son, Alexander.

After the war, Harvey and Alexander were among the first farmers in the area to explore and use new technology and methods. They "put a high priority on diversification of land use," according to their descendant Charles E. Harvey. The Harveys experimented with new pasture seed for grazing their cattle, and their results were closely evaluated and observed by other farmers and agricultural officials. New corn varieties were grown for their Hampshire hog operation and for horse and mule feed. A small flock of sheep was raised for wool and mutton, and some of the acreage produced oats and hay, though these later gave way to more cotton which was worked and harvested by tenants and sharecroppers. Through the decades, the Harveys have hosted students and farmers who have come to their farm to learn about progressive agricultural practices and more than 150 years of agrarian history.

THE TENNESSEE DEPARTMENT OF AGRICULTURE

The Tennessee Department of Agriculture evolved from the Bureau of Agriculture that was established by the Tennessee legislature in 1854 to promote agriculture through fairs and expositions. Several county fairs date their origins either before or just after 1865. One of the longest-running fairs in the state and in the entire south is the Gibson County Agricultural & Mechanical Fair that dates to October 1856. The fair disbanded from 1861 through 1868, but the county resumed this annual event in 1869 and has not missed a year since that time. The Bureau of Agriculture also ceased operating during the war, but Joseph B. Killebrew, a planter from Montgomery County, and other farmers advocated for its renewal after 1865 to encourage farming, industry, and commerce. When the legislature re-established the Bureau in 1871, Killebrew served as secretary of its six-member commission. Farmers across the state assisted Killebrew in collecting information. George E. Holman of Obion County was a reporter for the Bureau for six years. A native of Buckingham County, Virginia, Holman purchased the nearly 200-acre Holman Farm in Obion County in 1871. In addition to compiling data and operating his own farm, he was in the mercantile business in Union City.

In 1874, the agency released Resources of Tennessee, much of which Killebrew wrote. This landmark publication about the state's natural, agricultural, and industrial resources remains the authoritative study of this period. After exhaustive research, Killebrew reported that Tennessee farms and farmers were in "fairly good condition." He urged farmers to adopt "new South principles of scientific farming," which included breaking down large farms into small units that could be worked by an individual farmer or family; diversification; and the enrichment of soil through crop rotation, cover crops, and commercial fertilizers.

In 1875, the legislature abolished the Bureau and established the Bureau of Agriculture, Statistics, and Mines, for which Killebrew served as commissioner until 1881. The name of the agency was changed to the Department of Agriculture in 1893.

Lagoon Plantation
Haywood County

Property rights did not favor women in the nineteenth century, so few owned farms unless they inherited them from men. One exception was Mary Jaoqualine Smith Lee, whose husband, Col. Phillip S. Lee, III, was related to the family of "Light Horse Harry" Lee and his son, Robert E. Lee. After Col. Phillip Lee's death, Mary sold the family plantation in Campbell County, Virginia, and made an arduous journey to start a new life in Haywood County, Tennessee. With her two daughters and four grandchildren, slaves, and supplies, she travelled down the Atlantic Coast, crossed over to the Gulf of Mexico, and came up the Mississippi River. She purchased 500 acres in newly-formed Haywood County (1823) in 1828. By 1840, with the help

William Lee and Calista Anthony, with Ada Bailey and Benjamin Anthony Bailey, lived in the house built in 1865 by the veteran. The house is the residence of Bailey family today.

of her slaves which numbered sixty one men, women, and children, Mary Lee successfully managed and operated Lagoon Plantation. Corn, wheat, cotton, and livestock were the primary commodities, and she began to purchase additional land in Tennessee and Arkansas. The 1850 census lists her as owner of 18,000 acres in the two states.

By 1860, the Haywood County property had passed to her grandson, William Austin Anthony, born in Virginia in 1813. He was one of the largest planters in the county, relying on a slave population of about one hundred people to produce cotton crops, grains, and tend livestock. William had twelve children by his two wives. Milinda, who died in 1857, was the mother of ten children, and Julia Dyson, whom he married in 1858, gave birth to a son and a daughter. Both women were Moravians and against slavery. Anthony's children, including the females, were well educated at local schools and at academies in other states. When the Civil War began, two sons, William L. and Mark, were studying at the Moravian Male Academy in Lehigh Valley, Pennsylvania. They immediately returned to Tennessee. Mark joined the Union army and William enlisted with other local men in what would become Company D, 7th Tennessee Regiment (CSA). Family tradition maintains that the two sons enlisted in opposing armies as a deliberate act to ensure that one son would be on the victorious side when the war was over. The vagaries of war, however, resulted in the son on the winning side dying in battle. Mark Anthony died in Georgia on June 27, 1864, and is buried in the Marietta National Cemetery.

William Lee Anthony returned to his birthplace after surrendering in Alabama in 1865. That same year, he acquired about 200 acres of the original Lagoon Plantation and began construction on a house, about three miles from the two-story house built by his ancestor, Mary Smith Lee. In his obituary in 1925, this part of his life was described as a "crucial test" which Anthony met by returning to "manual labor to rebuild what had been destroyed and his splendid farm showed the results" of his efforts. He married Calista Taylor in 1880. With no children of their own, they raised siblings, Ada and Benjamin Bailey. The Anthonys adopted Benjamin and changed his name to Benjamin Anthony Bailey. He inherited Lagoon Plantation where the family continues to live in the 1865 house.

W. F. Pierce Farm
Obion County

Thomas Pierce bought 100 acres of wooded land in Obion County in 1851. When he died three years later, his eldest son, William Franklin (W. F.) Pierce, became the head of the household which included his mother, Caroline Moore Pierce, and six siblings. The family had a vegetable garden, an orchard, and livestock, and most likely grew cotton for a cash crop. W. F. Pierce joined the Confederate army in 1861 and spent four years with Company H, 47th Tennessee Infantry. He saw action at the major battles of Shiloh, Murfreesboro, Missionary Ridge, Chickamauga, Franklin, and Nashville. After being mustered out on May 17, 1865, Pierce returned to the damaged and neglected West Tennessee farmstead, bought out his siblings' shares, and began to plant crops and repair and rebuild fences and buildings. Working with Pierce was Essex, a former slave who, as a child, had been the property of Pierce's father, Thomas. The two men worked crops during the day and at night cleared more land for planting.

Pierce married Mary Jane Nooe in 1867, and they had three children. In 1873, the Paducah & Memphis Railroad (later the Illinois Central) came through the farm, and the town of Trimble was formed. In addition to farming, Pierce also operated a general merchandise store and held interest in two cotton gins. The veteran attended Confederate reunions and continued to be active in his community. He died in 1917 on the farm that is owned today by his great-great grandson, William Pierce Emge.

From left: W. F. Pierce, Mollie Pierce, Neil McKinnon, Lou Addie McKinnon, Willie Pierce, Thomas A. Pierce, and Jim McKinnon at the farmhouse in 1896.

German Catholic Homesteads
Lawrence County

During the winter of 1869-1870, representatives of the Cincinnati German Catholic Homestead Association came to southern Middle Tennessee to locate good and affordable farm land for resettling families displaced by the continuing conflict and repression in western Europe. The immigrants were dissatisfied with the crowded conditions in larger American cities and also the lack of jobs. The association purchased 15,000 acres in Lawrence County, and German and Polish families began to arrive. Eventually the settlers built four churches, a convent in Loretta and another in Lawrenceburg, and several schools. Most of the men were either tradesmen or skilled workers and had little farming experience. Nevertheless, the industry and determination of these families resulted in impressive agricultural enterprises that continue to be worked and owned by their descendants today.

Francis "Frank" Niedergeses, a native of Prussia and his wife, Sophia, founded the **Niedergeses Farm** in 1871. The couple left Cincinnati to move to 188 acres located at the head of Shoal Creek, a main artery that winds through the county, on land once lived on by frontiersman David Crockett. Here the couple raised their nine children. In addition to farming, the family also operated a tannery. Frank Niedergeses presided over the initial meeting of the first Catholic parish in Lawrenceburg, the county seat. The descendants of Frank and Sophia include the late James D. Niedergeses, the bishop of Nashville from 1975 to 1992.

The **Beuerlein Farm** was established in 1872 by Michael Beuerlein from Bavaria, who in 1870 had immigrated to Ohio, where he met his wife, Frances Renner. After the birth of their first child, they travelled by covered wagon to the Lawrence County property they acquired through the Homestead Association. This farm was part of the property initially purchased by Frank Niedergeses. A water supply came from a spring at the head of Shoal Creek, and the family still uses the same spring. The Beuerleins had seven children, though twin daughters died in infancy. Three of their sons were farmers and a fourth had a machine shop in the county seat of Lawrenceburg; their surviving daughter entered a convent. The Beuerlein family helped to build the first Sacred Heart Catholic Church, which later served as a school, in Lawrenceburg.

The Niedergeses family purchased their land in 1871. By 1890, their farmstead included well-ordered fields, a variety of livestock, good fences, and a farmhouse.
Courtesy *Lawrence County Archives*

This is a rare image of the first Sacred Heart Church in Lawrenceburg. It was built primarily by the first-generation German Catholic Homestead families in the 1880s. It was used as a school after the larger and current church was built.
Courtesy *Lawrence County Archives*

Nickolas and Anna Bauer Oehmen established their homestead north of Lawrenceburg in 1872. With a large family of fifteen children, the couple raised row crops along with cattle, swine, and poultry. They donated some of their land in the Ethridge community for a public school, known as Oehman School, which operated from 1880 until 1925. Their descendants own and manage the family farm as **Rocky Top Holstein Farm.**

Anna Bauer Oehman, who came to Lawrence County in 1871, is surrounded by her daughters in 1914.
Courtesy *Lawrence County Archives*

The Gang family was also part of the German migration to Tennessee. John Gang and his wife, Margaret, married in Ohio and followed other families south in hopes of building a better life. Gang purchased 232 acres north of Lawrenceburg on Buffalo Creek in 1878, and increased his holding by just over 40 acres in 1895. The family, including five children, raised corn, tobacco, horses, and cattle. Anton Gang acquired the family farm in 1899. He built a two-story house on the **Gang Farm** which is still lived in by his descendants more than a century later.

Anton Gang, second-generation owner of the 1878 farm, built this house in 1902. It is the primary residence of the descendants of the farm's founders who helped to rebuild the agricultural economy of the county after the Civil War.

During the last decades of the nineteenth century, farm families worked hard to build livestock herds. Horses and mules were raised and worked on the Frank Niedergeses Farm.
Courtesy *Lawrence County Archives*

END NOTES

Introduction

McKenzie, Robert Tracy. *One South or Many? Plantation Belt and Upcountry in Civil War-Era Tennessee* (Cambridge University Press, 1994).

Cleburne Jersey Farm

Wild, Amanda, and Carroll Van West. "Cleburne Jersey Farm, Maury County, Tennessee." National Register of Historic Places Nomination Form, Tennessee Historical Commission. Listed 2000.

Maplewood Farm

Bowman, Virginia McDaniel. *Historic Homes of Williamson County* (Nashville: Vanderbilt University Press, 1942; reprint, New York: Octagon Books, 1971).

Lowe, Karen E. *Antebellum Century Farms in Williamson County, Tennessee: Documents of Change and Continuity in Agricultural History*. Masters Thesis, (Murfreesboro, Middle Tennessee State University, 1999).

West, Carroll Van, "Historic Family Farms in Middle Tennessee Multiple Property Nomination." National Register of Historic Places Nomination Form, Tennessee Historical Commission. Listed 1994.

Brabson Ferry Plantation

Brabson, Estalena Rogers. *John Brabson I, Patriot of the American Revolution and Some of His Descendants* (Privately Published, 1975), 22-35.

Woodard Hall

West, Carroll Van, and Holly Rine. "Woodard Hall Farm (boundary increase and additional documentation) Robertson County, Tennessee." National Register of Historic Places Nomination Form, Tennessee Historical Commission. Listed 1994.

Insets

Post-War County Formation

Carpenter, J. W., and Michael Emrick. *Tennessee Courthouses* (London, KY: John W. Carpenter, 1996).

Combs, William H. "An Unamended State Constitution: The Tennessee Constitution of 1870," *American Political Science Review 32, no. 3* (June 1938): 514-524.

Hyde, Abigail R. "Lake County," in *Tennessee Encyclopedia of History and Culture,* http://tennesseeencyclopedia.net/entry.php?rec=759.

LaForge, Judy Bussell. "Tennessee's Constitutional Conventions of 1796, 1834, 1870: A Comparative Study," *West Tennessee Historical Society Papers 52* (1998): 64-80.

McLerra, Corinne. "Clay County," in *Tennessee Encyclopedia of History and Culture,* http://tennesseeencyclopedia.net/entry.php?rec=272.

Monroe, James. M. "James County," in *Tennessee Encyclopedia of History and Culture,* http://tennesseeencyclopedia.net/entry.php?rec=701.

Remembering Veterans

Butler, Margaret. *Legacy: Early Families of Giles County* (Pulaski, TN: Sain Publications, 1991).

Dargan, Elizabeth Paisley, ed. *The Civil War Diary of Martha Abernathy: Wife of Dr. Charles C. Abernathy of Pulaski, Tennessee* (Beltsville, MD: Professional Print, 1994).

The History of the Daughters of the Confederacy, Parts One and Two. (Kessinger Publishing, LLC, 2005).

Davies, Wallace E. "The Problem of Race Segregation in the Grand Army of the Republic." *Journal of Southern History 13, no. 3* (Aug., 1947): 354-372.

Losson, Christopher. "Grand Army of the Republic." *Tennessee Encyclopedia of History and Culture.* http://tennesseeencyclopedia.net/entry.php?rec=567.

Refer to the "Bibliographical Essay" for additional resource materials.

⋅≡ BIBLIOGRAPHICAL ESSAY ⋲⋅

No period of American history is as well-documented as the years 1861-1875, when the United States was divided and reunited. The 150th anniversary of the American Civil War, 2011-2015 has produced renewed interest, research and scholarship. This includes the identification and rediscovery of some sites and associated events, as well as the interpretation of places and personalities in the context of their enduring significance and legacy.

From the beginning of this project, family accounts in the Century Farm files were verified in primary and secondary sources whenever possible. Federal census records from 1850, 1860, and 1870; the Federal slave census from 1850 and 1860; and, for some farms, the 1860 and 1870 agricultural census provided detailed information and clarification. One of the most important primary sources for Civil War activity is *The War of the Rebellion: A Compilation of the Official Records of the Union and Confederate Armies*, published by the United States War Department between 1881 and 1901. These records are compiled in 127 volumes, plus a *General Index* and *The Official Military Atlas*.

Nomination forms for properties listed in the National Register of Historic Places are significant sources for describing the history, architecture, and context for several of the featured farms. County and community histories and compilations, as well as cemetery directories, genealogical studies, and extended family histories, add detail and verify names, dates, and relationships.

Former Confederate Gen. Marcus J. Wright's 1908 compilation, *Tennessee in the War, 1861–1865*, contains various lists of military organizations and officers from the state in both armies, including the Provisional Army of Tennessee. Among the tabulations are lists of Tennesseans in both Congresses and of the battles fought in Tennessee during the war.

In 1964, the state's Civil War Centennial Commission produced the two-volume *Tennesseans in The Civil War: A Military History of Confederate and Union Units with Available Rosters of Personnel*, which provides an authoritative, concise, and comprehensive history of each military unit that was organized in Tennessee during the war, as well as an alphabetical list, though not definitively complete, of Union and Confederate personnel.

Primary sources written by actual participants in the war provide valuable insight. The two-volume *Military Annals of Tennessee [Confederate]*, edited by Nashvillian John Berrien Lindsley and published in 1886, adds vivid and authentic first-hand accounts of wartime experiences to a collection of regimental histories and memorial rolls of each of Tennessee's Confederate military units.

Individual regimental histories written by the Civil War veterans contain much valuable information. Books such as *The Seventy-Ninth Highlanders: New York Volunteers in the War of the Rebellion, 1861-1865*, by William Todd, or former officers Samuel W. Scott and Samuel P. Angel's *History of the Thirteenth Regiment, Tennessee Volunteer Cavalry, U.S.A.*, utilize personal reminiscences to give an outsider's perspective to certain incidents recalled by some Century Farm family members.

Detailed personal stories about their service by more than 1,600 former soldiers are found in the *Tennessee Civil War Veterans Questionnaires*, available in manuscript form at the Tennessee State Library and Archives in Nashville, or in a five-volume set published by the Southern Historical Press, Inc., in 1985. This collection of thousands of unique documents gathered between 1914 and 1922 from men of all social classes provides needed veracity and particulars about wartime experiences.

In 1890, the state of Tennessee conducted a census of living Union Civil War veterans that asked for age, service history, and place of residence. The original records are available on microfilm; a published abstract, *1890 Civil War Veterans Census –Tennessee* (Evanston: Sistler and Associates, 1978) is also available.

Many Civil War veterans had been, or would become, community leaders and politicians. The abbreviated life histories of members of the Tennessee legislature can be found in the

Biographical Directory of the Tennessee General Assembly, primarily in Volumes I and II, published jointly by the Tennessee State Library and Archives and the Tennessee Historical Commission in 1975.

During 1886 and 1887, the Goodspeed Publishing Company issued a series of county histories under the general title of *A History of Tennessee from the Earliest Times to the Present*, which includes the biographies of prominent local men. In some instances, these accounts also give names of wives, mothers, and daughters, and often provide other helpful data such as military service, birth and death dates, and burial places.

The *Confederate Veteran* magazine published a large number of valuable personal memories of different aspects of the war. The series, edited until 1913 by founder Sumner A. Cunningham, was issued from 1893 to 1932. It contains first-hand accounts from general officers and private soldiers, as well as articles of general historical and genealogical interest.

In order to provide a larger social context for the individual stories that emerged from the chaos of civil conflict, the authors consulted the work of Stephen V. Ash. In *Middle Tennessee Society Transformed, 1860–1870: War and Peace in the Upper South* (Louisiana State University Press, 1988), Ash clearly demonstrates the breakdown of the hierarchical and communal bonds of southern society that led to the disintegration of the social order. Ash's study of the parts of the Confederacy that were taken over by the Union army during the war, *When the Yankees Came: Conflict and Chaos in the Occupied South, 1861 – 1865* (University of North Carolina, 1995), is useful in many ways, particularly in understanding the continuing resistance of Confederate sympathizers in the occupied areas.

Tennessee Farming, Tennessee Farmers: Antebellum Agriculture in the Upper South, by Donald L. Winters (University of Tennessee Press, 1994) is particularly useful in providing context for the state of farming in the geographical sections just prior to the Civil War. *One South or Many? Plantation Belt and Upcountry in Civil War-Era Tennessee*, by Robert Tracy McKenzie (Cambridge University Press, 1994) relies on primary source materials to compare plantations and smaller farms prior to and after the Civil War.

Mark Grimsley's *The Hard Hand of War: Union Military Policy Toward Southern Civilians, 1861–1865* (Cambridge University Press, 1995) gives an appreciation of the mindset and goals of the Union high command as they groped for a successful strategy for suppressing lethal guerilla activity without violating the rights of non-combatants. The initial phrase of the book's title, attributed to Gen. William T. Sherman, serves as a chapter heading in this publication.

ABOUT THE AUTHORS

Caneta Skelley Hankins and Michael Thomas Gavin collaborated on many projects, large and small, for over a decade. Major projects included the "Tennessee Iron Furnace Trail" – Web site, DVD, tours, symposium, and guide; "Creating a Regional Sense of Place" – survey, workshops, and heritage tourism planning guide for Hickman, Lewis, Perry, and Wayne counties; "Of Farms, Fences, and Furnaces: The Scots-Irish Influences on the Settlement Landscape of Tennessee" for the Ulster-American Heritage Symposium and its conference publication; and *Barns of Tennessee*, published by the Tennessee Electric Cooperative. Over the years, Hankins and Gavin enjoyed working together and with their colleagues at the Center for Historic Preservation, teaching students, assisting communities and groups, and continually learning about the historic resources of Tennessee. One from the South and one from the North, these friends and co-workers found common ground in their love of history, historic architecture, rock-and-roll music, family heritage, and Irish origins.

In 2007, Hankins and Gavin received a "Telly" award, in association with MTSU Audio & Visual Services, for their documentary film on the Tennessee Iron Furnace Trail.

Michael Thomas Gavin, a native of New Jersey, held a B. A. degree in English from Rutgers University and the M. A. degree in History from MTSU. For 25 years, he was president of a construction company that specialized in preserving and rehabilitating historic buildings. He spent much of his time identifying and restoring traditional houses and farm buildings throughout the state. Gavin's areas of research included African American history, the iron industry, and vernacular buildings. His survey and work on log buildings was a particular interest over many years. He authored or partnered on many reports and assessments of historic buildings and sites across the state, and his articles were published in various journals and books. He wrote several entries for the *Tennessee Encyclopedia of History and Culture,* and "Building with Wood, Brick, and Stone: Vernacular Architecture in Tennessee, 1770-1900," in *A History of Tennessee Arts: Creating Traditions, Expanding Horizons.* From 2002 until his death in 2013, Mike was the preservation specialist for the Tennessee Civil War National Heritage Area. He would likely dedicate this book to his wife, Linda, their children and grandchildren, his mother, and his many friends.

Caneta Skelley Hankins is a native Tennessean who grew up on a small but diverse family farm; she was among the last generation of southerners to pick cotton by hand. A graduate of Martin Methodist College in Pulaski, Tennessee, Hankins holds the B. A. in English and History and the M. A. in History with Emphasis in Historic Preservation from Middle Tennessee State University. Hankins was the projects coordinator of the Center for Historic Preservation from the time it was established in 1984 until 2001, when she was named assistant director, a position she held until her retirement in 2013. During that time, she was the director of the Tennessee Century Farms Program for twelve years. During her career of more than 35 years at MTSU, she was known for her national and international work in heritage education, Irish and Ulster-Scot resources in Tennessee, rural preservation, and for many publications and projects in varied areas of history and historic preservation. Among the seventh generation of her family to live in Williamson County, Hankins dedicates this book to the memory of her Skelley, Beasley, Grimes, and Walker great-grandparents who, thankfully, survived the Civil War in Tennessee.

INDEX

11th Illinois Infantry, 109

13th Indiana Cavalry, 28

14th Tennessee Cavalry, 91

16th Tennessee Infantry, 15

1870 Constitutional Convention, 146

18th Infantry, 116

1st Regiment of the Tennessee Volunteer Cavalry, 40

1st Tennessee Cavalry, 57, 69

1st Tennessee Infantry Regiment, 8, 132

1st Tennessee Partisan Ranger Regiment, 22

1st Tennessee Regiment, 132

2006 United States Survey of Saltpeter, 101

21st District of Williamson County, 125

22nd Tennessee Infantry, 90

24th Tennessee Infantry, 137

27th Tennessee Infantry Regiment, 20

2nd Kentucky Cavalry, 18

2nd Tennessee Infantry, 8

2nd West Tennessee Cavalry, 95

31st Infantry, Company D, 142

34th Virginia Cavalry, 47

3rd Tennessee Infantry, 132

3rd Tennessee Mounted Infantry, 41

3rd United States Colored Cavalry, 60

43rd Tennessee Mounted Infantry, 46

47th Tennessee Infantry, 157

4th Tennessee Cavalry, 91, 134

50th Tennessee Infantry, 18

5th Tennessee Cavalry, 33, 45

5th Tennessee Infantry, 19

61st Tennessee Mounted Infantry Regiment, 148

6th Tennessee Cavalry, 22

7th Tennessee Cavalry, 76

8th Tennessee Cavalry, 143

8th Tennessee Mounted Infantry, 70

8th Texas Cavalry Regiment, 16

9th Tennessee Cavalry, 56

A

Abernathy, Charles Clayton, Jr., 142

Abernathy, Martha, 142, 161

Adams, Alfred, 116

Adams, Amanda, 145

Adams, Col. John, 33

Adams, Reverend David, 145

Adams, Samuel, 145

Adams, Thomas, 145

African American, 114-115, 117-118, 119-122, 123-124, 127, 133, 165, 141

Alabama, 15-16, 27, 29, 48, 96, 101, 117, 139, 156

Albany, NY, 111

Aldridge, Rebecca, 95

Alexander Farm, 103-104, 111

Alexander Springs Farm, 96

Alexander, Absalom, 96

Alexander, Ellen Fields, 96

Alexander, James, 103

Alexander, Mack Keller, 96

Alexander, Mary, 103

Alexander, William Lawson, 103

Alford, Ben L., 127

Alford, Ben R., 127

Alford, Reverend Dr. Ben R., 127

Allen-Birdwell Farm, 4

Allen-White School, 121

Allen, Ella, 99

Allen, Fountain Pitts, 99

Allen, Bailey F., Jr., 90, 100

Allen, Laura M. Brown, 4

Allen, Bailey F., Sr., 99

Allen, James, Sr., 4

Allen, William B., Sr., 99-100

Allen's Bridge, 4

Allendale Farm, 99-100, 111

Allison Farm, 69

Allison, John, II, 69

Allison, Col. R.D., 138

Anderson County, 8-9

Anderson Farm, 35

Anderson, Charles, 152

Andersonville Prison, 56

Andersonville, Georgia, 56, 105

Andrew Tabernacle C.M.E. Church, 116

Anthony, Calista, 155

Anthony, Mark, 156

Anthony, Milinda, 156

Anthony, William Austin, 156

Anthony, William Lee, 155, 156

Appomattox, 3, 44, 142

Arkansas, 28, 87, 93, 101, 121, 145, 156

Armstrong, Frank, 138

Army of the Cumberland, 17

Army of the Tennessee, 60

Arno-College Grove, 125

Atlanta, 88, 152

Audobon, John J., 87

Austin Farm, 75

Austin, Charles Thomas Jefferson, 75

Austin, John Richard (J.R.), 75

Austin, Louisa Castellow, 75

B

Bailey, Ada, 155

Bailey, Benjamin Anthony, 155

Bailey, Ike, 145

Bailey, Joshua Curtis, 145

Bailiff, Columbus, 138

Bailiff, Eliza Foster, 138

Bailiff, James Monroe, 138

Ball's Farm, 91

Barnes, Will, 133

Barret Farm, 22, 24

Barret, Anthony Robert, 22

Barret, James M., 22

Barrett, Rebecca Hill, 22

Barretville Bank & Trust Company, 22

Barretville, 22, 24

Bate, Gen. William, 28

Battle of Atlanta, 152

Battle of Bean Station, 39

Battle of Chickamauga 157

Battle of Chickamauga, 143

Battle of Fort Donelson, 18

Battle of Fort Pillow, 91

Battle of Franklin, 157

Battle of Franklin, 30, 132

Battle of Hoover's Gap, 31

Battle of Missionary Ridge 157

Battle of Mossy Creek, 40

Battle of Murfreesboro, 157

Battle of Nashville, 157

Battle of Nashville, 29

Battle of New Hope Church, Georgia, 59

Battle of Perryville, 15

Battle of Seven Pines, 145

Battle of Shiloh, 157

Battle of Snow's Hill, 138

Battle of Stones River, 28, 33, 137

Battle of the Cedars, 28

Beachboard, Elizabeth, 31

Beachboard, Robert Walton, 31

Bean-Raulston Cemetery, 33

Beasley, Mary, 98

Bedford County, 17, 31, 123, 146

Bedford, Forrest, Gen. Nathan, 18, 20-21, 28, 42, 96, 138, 151

Beech Grove, 31

Beene, Capt. Robert, 32

Beene, Millie, 32

Bell Farm, 47, 63
Bell, David H., 47
Bell, Gen. Tyree H., 90
Bell, James, 47
Bell, Sarah, 47
Benton County, 81, 94, 150
Berry, Thomas, 97
Bethel Station, 95
Betsy, Aunt, 100
Beuerlein Farm, 158
Beuerlein, Michael, 158
Bible, Mary, 49
Big Bone Cave, 101
Big West Fork Creek, 99
Black Angus, 123
Black Family Land Trust, 115
Blackburn, Capt. James K. P., 16
Blackburn, Joseph H., 46
Blair, Albert, 36
Blair, James, 36
Blair, Kate, 36
Blair, Mary, 36-37
Blair, Rachel, 37
Blair, Wiley, 36
Blair's Ferry, 36
Bledsoe County, 8
Blockhouse #7, 28
Boden, James W., 19
Boden, Jeremiah, 19
Boden, Sarah, 19
Boggs Island, 143
Bond, Harriette, 124
Bond, Lucy Usher, 124
Bond, Nelson, 124
Boone, Daniel, 72
Boston, 98
Boyd's Creek, 143
Brabson Cemetery, 144
Brabson Ferry Plantation, 143-144, 160
Brabson, John, II, 143
Brabson, Benjamin Davis, 143
Brabson, Elizabeth, 143-144

Brabson, William Toole, 143
Bradford, Richard Stanford, 146
Bradley County 146
Bradley County, 48
Brady, Matthew, 59
Bragg, Gen. Braxton, 17, 27
Brame, Charles, 21
Branch, Carol, 151
Branch, Edward, 151
Branch, Tapp Craig, 118
Branner Cemetery, 40
Breckenridge, Maj. Gen John, 41
Bristol, 41
Brooks, Dillard, 92
Brooks, G.D., 92
Brooks, Tabor, 92
Brooks, William Calvin (W.C.), 92
Brown Guards, 132
Brown, Campbell, 132-133
Brown, Daniel, 134
Brown, Henrietta, 117
Browns Creek, 74
Brownsville, 124, 142
Buckingham County, 154
Buckner, Gen. Simon, 18
Buffalo Creek, 159
Bull's Gap, 41
Bullock Farm, 119
Bullock, Alice, 119
Bullock, Asbury, 73
Bullock, Frances Thomas, 119
Bullock, George, 119
Bullock, Henry, 119
Bullock, James, 119
Bullock, Maria, 119
Bullock, Mary Jane Robinson, 73
Bureau of Agriculture, Statistics and Mines, 154
Burrow-Gregory Farm, 71
Burrow, Elizabeth Ford Wilson, 71
Burrow, William S., 71

Butler Plantation, 152
Butler, Josiah, 122
Butler, Martha Lillard, 122
Butler, Perry, 122
Butler's Chapel, 122

C
Calfkiller River Valley, 46
Camden Road, 150
Camden-Decaturville Road, 150
Camden-Rockport Road, 150
Camp Douglas Prison, 106
Camp Douglas, 35, 56, 97, 105-106
Camp Morton, 143
Camp Sumter, 56, 105
Campbell County, 155
Campbell-Brown mansion, 132
Campbell, George Washington, 132
Campbell, McCoy "Mack", 132
Campbell, Sarah, 95
Caney Fork River, 137
Cannon, Capt. E. J., 40
Cardwell, Martha Washam, 148
Carroll County, 62
Carter County, 47, 61, 109
Cartwright-Russell Farm, 70
Cartwright, Amanda, 70
Cartwright, Clark, 70
Cartwright, Elizabeth, 70
Cartwright, Hailey, 70
Cartwright, Henrietta, 70
Cartwright, John, 70
Cartwright, Richardson Cloud, 70
Castellow, Benjamin, 75
Castellow, Henry D., 75
Castellow, Sarah Spruiell, 75
Cavalryman Jackson, 60
Cedar Creek, 51
Centralia, Illinois, 109
Century Farms, 36, 101, 115, 160, 165
Chancery Court, 119

Charles Hodges Thompson, 151
Charlotte, 55
Chattanooga University, 38
Chattanooga, 13-14, 17, 26-27, 32-33, 38, 48, 57, 63, 141, 146
Cheatham County, 59
Cherokee Nation, 36
Chester County, 146
Chester, North Carolina, 117
Chicago, Illinois, 56, 105
Chickamauga, 33, 103, 107, 138, 143, 157
Chicksaw Indians, 150
Cincinnati German Catholic Homestead Association, 158
Clark, Darius, 15
Clark, David, 15
Clark, Phineas, 15
Clarksburg, 62
Clarksville Weekly Chronicle, 141
Clarksville, 35, 99, 141
Clay County, 146, 161
Claybrook Cumberland Presbyterian Church, 152
Claybrook Road, 152
Cleburne Jersey Farm, 132-134, 160
Cleburne, Gen. Patrick R., 132
Cleveland, 48, 53, 141
Clifton, 20
Clinton, 8
Co. G., 4th Infantry, 75
Cocke County, 4
Coffee County, 31, 146
Collins, Daniel, 62
Collins, Miranda Lett, 62
Colored Methodist Episcopal Church, 152
Company D, 7th Tennessee Regiment, 156
Company F, 137
Company H, 1st Illinois Cavalry, 59
Company H, 47th Tennessee Infantry, 157

Confederacy, 2, 4, 33, 45, 52-53, 61, 87, 93, 101-102, 132, 138, 141-142, 147-148, 161, 163

Confederate Mound, 106

Confederate, 2-4, 6-8, 10-11, 13, 15-22, 26-30, 33-34, 39-42, 44-49, 53, 55-57, 59, 61-63, 67, 70, 72, 75, 86-88, 90-91, 93-97, 99-101, 103, 105-108, 111, 130, 132, 134, 137, 141-142, 144-145, 146, 148-149, 151, 157, 162-163

Connecticut, 134

Cook, Henry, 98

Cooper Farm, 149

Cooper, Isabella Dickson, 21

Cooper, James Irvin, 21

Copley, John M., 105

Corona Farm, 87, 93, 111

Country Wood Farm, 21

County Fermanagh, 5

Cox, John, 102

Cox, Robert, 152

Crack Hill Farm, 142

Craig, Amy, 118

Craig, Andrew, 118

Craig, Mary Jane, 118

Craig, McDonald, 118, 128

Craig, Tapp, 118

Craig, William, 118

Creek Wars, 32

Crockett County, 54, 91, 146

Crockett, David, 146

Crook, Elizabeth, 102

Crook, Gen. George, 16

Crossville, TN, 58, 63, 102, 111

Crumbley, Elizabeth Caroline King, 61

Crumley, James, 61

Cuff, Francis Asbury, 94

Cuff, Margaret, 94

Cuff, Sarah Sykes, 94

Cumberland County, 57-58, 63, 82, 101-102, 111

Cumberland Furnace, 115

Cumberland Mountains, 102

Cumberland Plateau, 45, 73, 149

Cumberland River, 116

Cupples, Douglas W., 53

Cypress Creek Farm, 81, 150

D

Dandridge, 40

Daniel, Joseph Crawford, 55

Daniel, Mary Loggins, 55

Daniel, William James, 55

Daniel's Dairy Farm, 55, 63

Daugherty, Susan, 92

Davidson County, 29

Davis Farm, 74

Davis, Andrew Hunter, 116

Davis, Annie Wood Wilkinson, 74

Davis, Arista Hare, 74

Davis, Asa, 74

Davis, Bunkum, 116

Davis, Columbus, 74

Davis, Emma, 116

Davis, James Harvey, 116

Davis, Jefferson, 18

Davis, John E., 4, 143

Davis, John, 116

Davis, Martha Jane Breazeale, 74

Davis, Robert S., 46

Decatur County, 20, 95, 97

DeKalb County, 138

Devil's Elbow, 87, 93, 111

Dickson County, 55, 63, 97, 114-115, 146

Dickson, Jane Moore, 21

Dillard Brooks Farm, 92

Divine, Nathan, 119

Divine, Sallie, 119

Divine, Susan, 119

Dixie Farm, 34

Dixontown, 117

Donelson, John, 70

Dover Hotel, 18, 84

Dr. Keeton, 20

Drake Farm, 119

Drake, Penelope, 116

Dry Creek, 138

Duck River Furnace, 134

Duck River, 17, 134

Dunbar, 20

Duplex, 136

Dutch, 10

Dyer County, 54, 88, 90, 92

Dyersburg, 88

Dyson, Julia, 156

E

E.A. Cuff Farm, 94

Earnest, Benjamin Franklin, 108

East Tennessee and Georgia Railroad, 10

East Tennessee and Virginia Railroad, 41

East Tennessee Mounted Infantry, 6

Easterly, Anna Parrott, 49

Easterly Farm, 49

Easterly, Francis Marion, 49

Easterly, Frank, 49

Easterly, Jacob, 49

Easterly, John George, 49

Easterly, Mary Harpine, 49

Easterly, Matilda Robeson, 49

Easterly, Narcissa Powell, 49

Eaton, Chaplain John Eaton, Jr., 20

Edgefield and Kentucky Railroad, 139

Edgman Farm, 19

Elk River, 117

Elkton, 117

Elliott, Gen. Washington L., 4

Ellis, Capt. Daniel "Old Red Fox", 109

Elmwood Farm, 42, 83, 108

Elmwood Plantation, 27

Emancipation Proclamation, 9

Emge, William Pierce, 157

Episcopal Church, 1896, 148

Erwin, Mrs. Alexander E., 142

Essex, 157

Estanaula Oaks, 158

Ethridge, 159

Ewell Farm, 133

Ewell, Gen. Richard Henry, 132

Ewell, Lizinka Campbell Brown, 132

F

Fairview Farm, 40, 84

Farmington, 16, 122

Farnsworth, Benjamin, 106

Farnsworth, Elizabeth Parman, 105, 106

Farnsworth, John W., 105

Fayette County, 21, 153

Federal, 4, 8, 10-11, 13-16, 19-22, 26, 28-30, 34, 37, 39-40, 45-47, 51-53, 86-87, 91, 93, 99-101, 104, 106, 114, 120, 122, 130, 132, 134, 138, 141, 144, 147, 152, 162

Fentress County, 149

Fermanagh-Ross Farm, 5, 24, 80

Flatwoods Methodist Church, 94

Flowers Branch Cemetery, 51

Ford Motor Company, 123

Ford, Christopher A., 57-58

Ford, John, 57-58

Ford, Dr. John, Jr., 57

Ford, Mary Jane, 57

Ford, Nancy Loden, 57

Forrest, Gen. Nathan B., 96

Fort Donelson National Military Park, 18

Fort Donelson, 18, 26, 35, 99

Fort Hill Cemetery, 48

Fort Nashborough, 70

Fort Pickering, 87, 93

Fortress Rosecrans, 28, 122

Foulks (Hutchinson) Farm, 77

Foulks, Elizabeth, 77

Foulks, John J., 77

Fowler-Lenoir Farm, 11-12, 24

Fowler, Cora, 11

Fowler, Mary Josephine Kelso, 11

Fowler, William J., 11

Franklin County 146

Franklin Turnpike Company, 29

Franklin, 3, 26, 29-30, 97, 108-109, 125, 132, 142, 146, 157

Freedman's Aid Society, 148

Freedmen's Bureau, 122

French Broad River, 143

Friendship, 88

Ft. Donelson, 3, 18, 55, 59

Ft. Henry, 59

G

Galbraith, Lt. Col. Robert, 33

Gallatin, 90, 119

Gang Farm, 159

Gang, Anton, 159

Gang, John, 159

Gang, Margaret, 159

Gardner, Alexander, 59

Gardner, Henrietta Brown, 117

Gardner, Martin, 117

Gardner, Matt, 79, 117, 128

Gardner, Rachel, 117

Gass, Margaret "Peggy", 4

Gauldin Farm, 88, 90

Gauldin, Margaret, 90

Gay, Capt. William, 88

Gehmen, Nickolas, 159

Gen. Johnston, 137

Georgia Partisan Rangers, 37

German Catholic Homesteads, 158

German, 10, 158-159

Gettysburg, 132

Gibson County Agricultural and Mechanical Fair, 154

Gibson County, 21, 154

Giles County Medical Society, 142

Giles County, 16, 79, 117, 128, 142

Gillem, Brig. Gen. Alvan, 41

Gillespie, James W., 46

Glen Villa, 143

Goddard Mountain Farm I, 101

Gordon Farm, 52, 63

Gordon, A.W., 52

Gordon, Amanda Nelson, 52

Gordon, Gilbert, 52

Gordon, Ginny, 52

Gordon, John Hilton, 52

Gordon, Robert Jennings, Sr., 52

Grainger County 39, 146

Grand Army of the Republic, 141-142, 161

Grand Junction, 120, 128

Grant Memorial University, 38

Grant, Gen. Ulysses S., 18, 128, 141

Grassy Cove Methodist Church Cemetery, 58

Grassy Cove, 57-58, 63, 82, 101-102

Graves, Charles David, 148

Graves, Henry, 125

Graves, Reverend William C., 148

Gray, Federick, 18

Gray, Rachel Fulkerson, 18

Green, John U., 22

Greene County, 4-5, 24, 49, 80, 105-106, 108

Greeneville, 6, 141

Gregory, Jack Burrow, 71

Grierson, Col. Benjamin, 22

Guthrie Farm, 118

Guthrie, Andrew, 118

H

H.E.F. Blair and Sam Blair Farms, 36

Hall, Maden, 5-6

Hamblen County, 40, 146-148

Hamilton County, 146

Hamilton, Ida, 29

Hammons, Dillia Ann, 153

Hampton, Lt. Gen. Wade, 138

Hardeman County, 120-121

Harris Station, 77

Harris, Gov. Isham G., 101

Harrodsburg, Kentucky, 98

Harvey, Charles E., 153

Harvey, Dr. Richard Henry, 153

Hatchie River, 22

Hawkins County, 27, 41

Hawkins, Harriet Smith, 88, 90

Hawkins, Isabella Taylor, 88, 90

Hawkins, Simon Peter, 88, 90

Hawkinsville, 90

Hayes, President Rutherford B., 130

Haywood County, 124, 142, 155-156

Henderson County, 74, 78, 152

Henderson, Isaac, 122

Henderson, Lavinia, 122

Henderson, Judge Logan, 122

Hendrix Farm, 57, 63

Henning, 76

Henry County, 19

Hester Farm, 21

Hester, Elizabeth Brame, 21

Hester, John, 21

Hickory Valley, 73

Hill-Johnson Cemetery, 95, 97

Hill, Harvey, 153

Hill, Holly, 98, 111

Hill, John Prior, 95

Hill, Mary Elizabeth, 95

Hilltop, 153

History of Dyer County, 90

Holley, George, 110

Holley, Margaret S., 110

Holman Farm, 154

Holman, George E., 154

Holt Farm, 21

Homestead Association, 158

Hood, Gen. John Bell, 28, 30, 96

Hord, Amelia, 27, 28

Hord, Thomas, 28

Horner Farm, 51, 63

Horner, Nancy Randel, 51

Horner, William, 51

Hoskins, John Gaston, 35

Hoskins, Neander, 35

Hotchkiss Valley Road, 104

Houston County, 146

Huff's Ferry, 38

Humboldt, 53

Humphreys County 146

Hunter Cove Farm, 72

Hunter, Amy, 72

Hunter, Dudley, 72

Hunter, Rush, 72

Hunter, Sarah Boone, 72

Hunter, Vance, 72

Hunter, William, 72

Hunter's Point, 116

Hurst, Col. Fielding, 22, 92

Hutchinson, Martha Allen, 77

I

Illinois Central, 157

Illinois, 35, 56, 59, 97, 105, 109, 141, 157

Indian Corn, 134, 139

Ingram, Preston, 98

Ireland, 5, 47

Irish, 5, 10, 38, 90, 139, 164-165

Irwinsville, Georgia, 18

Isaac Huddleston Farm, 73

Island #37, 87

J

J & J Farm, 11-12

J.M. Bailiff Farm, 138

Jackson County, 146

Jackson, Gen. Andrew, 96

Jacobs Farm, 31

Jacobs, Alfred, 31

Jacobs, Catherine Dillard, 31

Jacobs, Jeremiah, 31

Jacobs, Rebecca Rudd, 31

James County, 146, 161

James, Lissie Reed, 40

Jasper-Sewanee Road, 32-33

Jefferson City, 40

Jefferson County, 40, 56, 84, 146

Jenny-Ben Farm, 105

Jersey Bulletin and Dairy World, 113, 133

Jim Crow laws, 115, 117

Joe Carter Farm, 146

John C. Kemmer Farm, 102

Johnson City, 61, 110

Johnson County, 7

Johnson Farm, 142

Johnson, Andrew, 53, 132

Johnson, Frannie Elizabeth, 95

Johnson, James, 95, 97

Johnsonville, 151

Jones Creek, 97

Jones, J.S., 148

Jonesboro, Georgia, 142

Jonesborough, 69

Julius Rosenwald Foundation, 121

K

Keeton Farm, 20

Keeton, Catherine, 20

Keeton, John Lawson, 20

Keller, Hiram Washington, 76

Keller, Roberta Burks, 76

Kelso, Charles, 11

Kelso, Elizabeth Wyley, 11

Kelso, James Wyley, 11

Kemmer, Andrew, 102

Kentucky, 8, 15, 18, 27, 47-48, 57-58, 70, 72, 98-99, 101, 109, 119, 138-139, 149

Killebrew, Joseph B., 154

Kimbro, Allen, 17

Kimbro, George, 17

Kimbro, Mary Elkins, 17

Kimbro, Thomas, 17

Kimbrough, John Payton, 19

Kimbrough, Nancy Higginson, 18

King, Alfred J., 61

King, William B., 61

Knight Farm, 31

Knob Creek Museum, 110

Knob Creek, 61, 63, 110

Knox County, 145

Knoxville, 24, 26, 41-42, 61, 64, 103, 111, 137, 144

Krouse, Anna, 110

Krouse, Daniel, 109

Krouse, Susannah Wine, 109-110

L

L Co, 8th Tennessee Cavalry, 143

Ladies Hospital Association, 132

Lagoon Plantation, 155-156

Laguardo, 116

Laird, Mackie, 16

Laird, Nancy, 16

Laird, Robert, 16

Lairdland Farm, 16

Lake County, 146, 161

Lamar Farm, 8

Lamar, Joseph B., 8

Lamar, Nancy Wallace, 8-9

Lamar, William, 8-9

Lancaster Farm, 137

Lancaster, Elizabeth, 137

Lancaster, John, Jr., 137

Lancaster, William, 137

Land Trust for Tennessee, 98

Landrum, Beulah, 123

Landrum, Cora McLain, 123

Landrum, Genie, 123

Landrum, Jesse, 123

Lane, Thompson, 150

Lanier, Charles, 123

Lauderdale County, 75-76

Lawrence County, 96, 158-160

Lawrenceburg, 96, 158-159

Lee, Charles, 134

Lee, Elizabeth Brown, 134

Lee, Col. Phillip S., III, 155

Lee, John N., 136

Lee, John Wills Napier, 136

Lee, Samuel Brown, Jr., 134

Lee, Mary Jacquline Smith, 156

Lee, Robert E., 155

Lee, Samuel Brown, 134

Lee, Susan Amanda Napier, 134

Leipers Creek, 98

Leven, Glen, 29-30, 42

Levi Trewhitt Farm, 48

Lewisburg, 16

Liberia, 34

Liberty Gap, 31

Lick Creek, 118

"Light Horse Harry" Lee, 155

Light House Farm, 88

Light, Joel A., 88

Light, Sue, 88

Limbro, Benjamin S., 17

Limestone Cove, 47, 63

Lincoln County 146

Lincoln, Abraham, 2, 9, 59, 87, 114, 143

Little Tennessee River, 11

Liverpool, England, 34

Loggins, Gen. John Littleton, 53, 55, 63

Loggins, James, 55

Loggins, Katherine Rebecca, 55

Loggins, Nancy Grimes, 55

Loggins, Thomas, 55

Lone Pine Farm, 61, 63

Longmire, Rose Lamar, 8

Longstreet, Lieut. Gen. James, 38

Lookout Mountain Cave, 101

Looney, Joel, 14

Looney, Robert, 14

Looper-Thompson Farm, 34

Looper, Joseph, 34

Looper, William, 34

Lost Sea, 101

Loudon County, 36, 38, 80, 103-104, 126, 146

Loudon High School, 38

Loudon, 36-38, 80, 103-104, 126, 141, 146

Louisiana, 24, 28, 63, 139, 163

Louisville, KY, 72, 90, 138

Love, Aline Boden, 19

Luster, Alex, 125

Luster, Anna, 125

Luster, Anthony W., 125

Luster, Betsey, 125

Luster, Grant, Jr., 125

Luster, Jennie, 125

Luster, Mattie, 125

Luster, Nelson, 125

Luster, Sallie Jones, 125

Luster, Grant, Sr., 125

M

MacArthur, Maj. Arthur, 4

Macedonia, 24, 119

Macon County, 71

Madison County, 152

Madisonville, 11

Malone, Frances E. Drake, 119

Malone, Richard, 119

Manchester, 31

Maneys of Oaklands, 122

Mann, Dick, 124

Mann, Harriet Taylor, 142

Mann, P.H., 124

Mann, Richard A., 142

Manscoe's Creek and Springfield Turnpike, 139

Mansell Farm, 78

Mansell, George Washington, 78

Maplewood Farm, 134, 136, 160

Marietta National Cemetery, 156

Marion County, 13, 32-33, 82, 101

Martin, Gen. William, 40

Martin, TN, 92

Mashburn, William, 107

Mason Farm, 38

Mason, Eliza Kerr, 38

Mason, Elizabeth "Bettie", 38

Mason, Mary Jane, 38

Mason, Thomas Jefferson, 38

Massengill Farm, 39, 42

Massengill, Michael, 39

Massengill's Mill, 39

Maury County, 79, 91, 132, 160

Mayo, David E., 91

McCullough, Benjamin, 88

McCullough, Emily, 41

McDonald Craig Farm, 118, 128

McKenzie, Robert Tracy, 163

McKinnon, Jim, 157

McKinnon, Lou Addie, 157

McKinnon, Neil, 157

McKnight, Caroline, 122

McMillan-Alexander Stagecoach, 96

McNairy County, 95

McPeak Farm, 15

McPeak, John Lambert, 15

McPeak, Mahaley Clark, 15

McQueen Farm, 38, 80

McQueen, Edmund Preston, 38

Meachum, Barns, 152

Meadow Dale Farm, 17

Memphis & Charleston Railroad, 21

Memphis, 3, 21-22, 24, 53, 63, 78, 87, 91, 93, 121, 134, 141, 157

Methodist Episcopal Church, 148

Mexican War, 88, 132

Michael Krouse Farm, 59, 109, 111

Middle Tennessee, 3, 16, 18, 24, 29, 31, 42, 63, 67, 128, 158, 160, 163, 165

Military Board, 101

Military Road, 96

Miller Family Cemetery, 61

Miller, Alfred, 61

Miller, George W., 61

Miller, Lafayette, 61

Milroy, Maj. Gen. R.H., 28

Minutes of the Holston Conference of the Methodist, 148

Missionary Ridge, 137

Mississippi River, 3, 86-88, 93, 155

Mississippi Valley, 120

Mississippi, 3, 21, 29, 35, 86-88, 93, 98, 116, 120, 134, 139, 155

Missouri mules, 72

Mobile and Ohio Railroad, 21

Mobile, Alabama, 15

Monroe County, 11-12, 101

Montgomery County, 35, 99, 111, 119, 154

Moore County, 146

Moore Farm, 41-42

Moore, Howard G., 41

Moore, John Rufus, 41

Moravian Male Academy, 156

Moravians, 156

Morganton, 12

Morrell, E.S., 69

Morrell, Susan Allison, 69

Morristown Normal Seminary, 148

Morristown, 61, 148

Mount Farm, 54

Mount, Harris, 54

Mousetail Landing, 118

Mt. Pisgah Methodist Church, 88

Mt. Sterling, Kentucky, 119

Murfreesboro, 19, 27-28, 31, 42, 52-53, 122, 128, 157, 160

N

Napier, John Wills, 136

Napier, Susan Amanda, 134

Nashville and Alabama Railroad, 29

Nashville and Chattanooga Railroad, 17, 27

Nashville and Knoxville Railroad, 137

Nashville, 3, 17, 24, 26-30, 33-34, 42, 53, 67, 70, 90, 96, 101, 111, 121, 132, 137, 139, 141, 147, 157-158, 160, 162

National Encampment, 141

National Register of Historic Places, 4, 24, 28, 100, 108, 111, 117-118, 132, 136, 160-162

Native Americans, 127

Negley, Gen. James, 33

New Orleans and Ohio Railroad, 77

New Orleans, 77, 87, 96

New Town community, 123

New York, 63-64, 103, 111, 139, 160, 162

Newman Farm, 56

Newman, Jane Caldwell, 55

Newman, John "Black Jack", 56

Newman, Jonathon, 56

Nickajack Cave, 101

Niedergeses Farm, 158, 160

Niedergeses, Francis "Frank", 158

Niedergeses, James D., 158

Niedergeses, Sophia, 158

Nolichucky River, 4, 108

Nooe, Mary Jane, 157

Normandy, 17

North Carolina, 49, 52, 75, 91, 105, 125, 143, 163

Nunn, Lucy Ann, 91

Nunn, David, 91

Nunn, Elsey, 91

O

O'Sullivan, Timothy, 59

Oak Woods Cemetery, 105

Oakdale Farm, 59

Oakview Baptist Church, 124

Oath of Allegiance, 45, 53-54

Obion County, 77, 146, 154, 157

Obion River, 62

Oehmen, Anna Bauer, 159

Officer Cemetery, 45

Officer Farm, 45, 63

Officer, Abraham H., 46

Officer, John, 46

Officer, William A., 45

Ohio River, 87, 93

Old Kentucky Road, 15, 72

Old Ridge Road, 5

Old Woodbury Highway, 122

Osburn, Mary Jane, 99

Ottinger, Jacob, 107

Overton County, 34, 45, 63, 146

P

Paducah and Memphis Railroad, 157

Paris, 19

Parker's Crossroads, 152

Parman, Elizabeth, 105

Parrottsville Academy, 49

Parrottsville, 4, 49

Pearson, Eliza, 152

Pearson, John Solomon, 52

Pearson, Jonathan Dudley, 152

Pearson, Little Billy, 152

Pearson, Susan Elizabeth, 152

Pearson, William, 152

Perry County, 51, 63, 118

Perryville, Kentucky, 15, 98

Phelps, Doris Tarwater, 10

Pickett County, 146

Pierce, Caroline Moore, 157

Pierce, Mollie, 157

Pierce, Thomas A., 157

Pierce, William, 157

Pigeon Forge, 101

Pikeville, 141

Pioneer Homestead Farm, 110

Planter's Bank, 87

Plemmons, Douglas, 101

Possum Hollow, 138

Powder Mill Cave, 101

Presnell, Alexander M., 38

Prichard Cemetery, 137

Prichard, James C., 137

Prichard, Melissa, 137

Primitive Baptist Congregation, 117

Promise Land, 114

Pulaski Medical Society, 142

Pulaski, TN, 16, 58, 92-93, 141-142, 161, 165

Putnam County, 72-73, 78

Q

Quinn, John M., 95

R

Radical Reconstruction, 141

Ragon, Horace, 12

Ragon, Jane Lilliard, 12

Ragon, Joseph Charles, 12
Randel, Amos, 51
Randel, Rebecca, 51
Raulston, Col. James, 32
Raulston, Sam Houston, 33
Raulston, William Henry, 33
Reconstruction, 63, 99, 110, 115, 130, 141
Reelfoot Lake, 146
Reese, Mary Porter, 40
Religious Intelligencer, 148
Renner, Frances, 158
Resources of Tennessee, 164
Rest Haven Cemetery, 142
Revelle, Corday B., 91
Rheatown, 5
Rhodes Farm, 95
Rhodes, Claudine, 95, 97
Rhodes, Janice, 107
Rhodes, Leon, 107
Rice Farm, 53
Rice, William C., 53
Richmond, VA, 132, 137, 145
Rising Sun Farm, 116
Roane County, 38
Robertson County, 81, 127, 139, 161
Robertson Farm, 120, 128
Robertson, Crawford, 121
Robertson, Evelyn, Sr., 121
Robertson, Jerry, 114
Robertson, Evelyn C., Jr., 121
Robertson, Mae Jane, 114
Robertson, Myrtle, 121
Robinson, James S., 73
Robinson, Syrena, 73
Rock Island, 137
Rocky Top Holstein Farm, 159
Rogers, Alexander, 153
Rogers, Mary Jane, 153
Rogersville Junction, 41
Rosecrans, Gen. William S., 17
Rosenwald, Julius, 117, 121

Ross, William, II, 5
Ross, William, III, 5
Ross Road, 5
Ross, John Gass, 5
Ross, Mary Elizabeth, 6
Ross, Vincent, 6
Rover, 141
Rucker, 52
Russell, Leonidas (Lon) Shelby, 70
Russellville, 40, 61
Russia, 132
Rutherford County, 27, 42, 52, 63, 83, 122-123
Rutledge, 39-40

S

Sacred Heart Catholic Church 158-159
Samuel Raulston Farm, 32, 82
Sander, John J., 97
Sanders Spring Forest Farm, 97
Sanders, Henry, 97
Sanders, John, 97
Sanders, Susan West, 97
Scales, Everlee Lanier, 123
Scales, Maxine, 123
Scales, Tony, 123
Seaton, William P., 4
Sequatchie River, 13-14
Sevier County, 10, 143-144
Shady Oaks Farm, 148
Shelby County, 22, 24
Sherfy, David Preston, 59, 60, 64, 109-111
Sherfy, Isabell Krouse, 109-110
Sherfy, John, 110
Sherman, Gen. William T., 12
Shiloh, 26, 59, 109, 142, 157
Shoal Creek, 158
Shropshire, Lt. Francis C., 37
Signal Mountain Road, 13
Simon, 18, 88, 90
Skelley Farm, 98, 111

Skelley, Henry Potts, 98
Skelley, James Crawford, 98
Skelley, Sarah Louise Potts, 98
Smith County, 70, 137
Smith Farm, 88
Smith, J.W., 88
Smith, William H., 59
Somerville, 153
South Carolina, 51, 94, 128, 137
South Pittsburg, 32
Southern Claims Commission, 5
Southern Cross of Honor, 142
Sparkman-Skelley Farm, 98, 111
Sparta, 15, 45, 72
Spears, Gen. James G., 8
Spring Hill, 132, 134
Spring Meadow Farm, 145
Stewart County, 18, 84, 146
Stewart, Gen. Alexander P., 31
Still Hollow Farm, 4, 24
Stockley, Charles Ambrose, 87, 93
Stokes, Col. William B., 33
Stones River National Cemetery, 114
Strawberry Plains, 6, 40, 145
Sugartree Farm, 127
Sullivan County, 53
Sumner County, 119
Surrender House, 18, 84
Swan, Elizabeth, 57-58
Swan, Robert, 58
Sweeten's Cove, 32-33
Sweetwater, TN, 101

T

Tarwater Farm, 10, 24
Tarwater, Charles B., 10
Tarwater, Matthew, 10
Tarwater, Sarah Rule, 10
Taylor, Gen. Zachary, 88
Taylor, Henderson, 151
Teague Farm, 13-14
Teague, Elijah, 13

Teague, Joel E., 13
Teague, Rebecca Looney, 14
Tennessee 49th Infantry, 35
Tennessee and Georgia Railroad, 147
Tennessee and Virginia Railroad, 145
Tennessee Department of Agriculture, 94, 154
Tennessee General Assembly, 24, 146
Tennessee Legislature, 12, 30, 139, 154, 162
Tennessee River, 12, 19-20, 36-38, 51, 94, 118, 150-151
Tennessee State Medical Society, 142
Tennessee State University Tennessee Agricultural & Industrial State Normal School, 121
Tennessee, 2-12, 15-20, 22, 24, 26-31, 33-42, 44-49, 51-53, 55-57, 60-63, 67, 69-70, 75-76, 86, 88, 90-96, 98-99, 101, 103, 107-109, 111, 114-115, 117-121, 127-128, 130-134, 137, 139, 141-148, 150-151, 154-165
Terry's Texas Rangers, 16
The Grange, 94
The Memphis Daily Appeal, 141
The Ward Farm, 101
Thirteenth Amendment, 114
Thomas, Eliza, 116
Thomas, Gen. George, 30
Thompson, Arena Jane, 151
Thompson, Charles Hodges, 150
Thompson, Dave, 150-151
Thompson, Jerry, 150
Thompson, John, 29-30
Thompson, John M., Jr., 30
Thompson, Mary, 29-30
Thompson, Nancy "Velonia", 151
Thompson, Penelope, 150
Thompson, William, 150
Thruston, Gen. Gate P., 29
Tilson Farm, 83, 107
Tilson, Capt. William E., 107
Tilson, Catherine, 107

Tilson, Elizabeth Beals, 107
Tilson, George, 107
Tilson, James W., 107
Tilson, Marion, 107
Tipton County, 24, 87, 93
Todd, William, 103, 162
Toole, Elizabeth Berry, 143
Trail of Tears, 38
Trenton, 21, 88, 95
Trewhitt, Andrew J., 48
Trewhitt, Daniel, 48
Trewhitt, Levi, Jr., 48
Trewhitt, Levi, Sr., 48
Trewitt, Harriett Lavendar, 48
Trigg, Elizabeth Bradley, 87
Trigg, John, 87
Trimble, 157
Trousdale County, 146
Tullahoma Campaign, 17, 31

U

U.S. Constitution, 114
U.S. Hwy. 43, 96
Unicoi County, 47, 83, 107, 146
Union and Confederate forces, 145
Union Army, 14, 27-28, 45, 61, 66, 86, 88, 156, 163
Union City, 154
Union, 2-5, 7-8, 10, 14-15, 17, 19, 21-22, 26-30, 33-34, 37, 39-41, 44-46, 48-49, 52-53, 55-57, 59, 61, 63, 66, 69-73, 75, 86-88, 91, 93, 95-96, 102-105, 108-109, 114, 116, 122, 130, 141, 145, 147-149, 151, 154, 156, 162-163
United Confederate Veterans, 106
United Daughters of the Confederacy, 141
United States Colored Troops, 30, 34, 128
Upper Cumberland Plateau, 45, 149
Utley, Fredonia "Adaline", 150-151
Utley, Penelope. 151

V

Van Buren County, 101
Vance, Laura Blair, 36
Vaughn Farm, 18, 84

Vicksburg, Mississippi, 35, 98
Virginia, 5, 22, 26, 41, 47, 49, 56-57, 62, 69, 105, 125, 132, 137, 143, 145, 154-156, 160

W

W.F. Collins Farm, 62
W.F. Pierce Farm, 157
Wagner-Worley Farm, 7
Wagner, David, 7
Wagner, Margaret Weitzel, 7
Wagner, Mary Catherine Hagey, 7
Wagner, Nathaniel T., 7
Walden's Ridge, 13-14
Walker, Ann Wiley, 34
Walker, John Alphonso, 34
Walnut Flat Farm, 152
War of 1812, 32, 74
Wartrace, 31
Washington County, 59, 61, 63, 69, 109
Washington, Foster, 127
Washington, Joseph, 127
Washington, Lawson, 127
Washington, Marina, 127
Washington, President George, 127
Waters, Shelah, 46
Weakley County, 67, 92-93
Weaks, Andrew, 18
Weaks, James, 18
Weaks, Nancy Gray, 18
Weaks, William B., 18
Webster, William J., 133
Wessyngton Plantation, 127
West Tennessee, 3, 20, 22, 53, 62-63, 75, 90-91, 94, 111, 120, 146, 150, 161
West, Susan Thompson, 30
Western Highland Rim, 118
Wheeler, Gen. Joseph, 13
White County, 15, 34, 73
White Family Cemetery, 61
White, Daniel, 147
White, Jimeson, 147
White, John, 147
White, Jonah, 147

White, Joseph, 147
White, Washington, 147
White, William, 147
Whitetown Acres Farm, 147
Whiteville, 121
Wieber, Susan Austin, 75
Wilberforce College, 119
Wilder, Col. John T., 31
Wiley Woodard and Company, 139
Williams, Isaac, 122
Williams, Stokley Donaldson, 40
Williamson County, 98, 111, 125, 132, 134, 160, 165
Willis, Sampson, 152
Wilson County, 53, 116
Wilson, Bell, 5
Winchester, 144
Wisdom County, 146
Wisha, Seabo, 137
Witcher, Col. V.A., Jr., 47
Wolf Creek, 138
Woodard Hall, 81, 139, 161
Woodard, Thomas, 139
Woodard, Wiley, 139
Woodlawn Baptist Church, 124
Woodlawn Cemetery, 30
Woolfolk, John George, 152
Woolson, Albert, 141
Wooten-Kimbro Farm, 17
World War II, 4, 126
World's Championship Pacing Record, 136
Worley, Mary Ann, 7
Worley, Tom, 7
Wright, America, 149
Wright, Jacob, 149

Y

Yankeetown, 34
Young, Samuel, 118

Z

Zion Hill Baptist Church, 88